Agrarian Reform in Reverse
The Food Crisis in the Third World

Agrarian Reform in Reverse
The Food Crisis in the Third World

edited by
Birol A. Yesilada, Charles D. Brockett,
and Bruce Drury

Westview Press / Boulder and London

Westview Special Studies in Agriculture Science and Policy

This Westview softcover edition is printed on acid-free paper and bound in softcovers that carry the highest rating of the National Association of State Textbook Administrators, in consultation with the Association of American Publishers and the Book Manufacturers' Institute.

All rights reserved. No part of this publication may be reproduced or transmitted in any form or by any means, electronic or mechanical, including photocopy, recording, or any information storage and retrieval system, without permission in writing from the publisher.

Copyright © 1987 by Westview Press, Inc.

Published in 1987 in the United States of America by Westview Press, Inc.; Frederick A. Praeger, Publisher; 5500 Central Avenue, Boulder, Colorado 80301

Library of Congress Catalog Card Number: 87-29549
ISBN: 0-8133-7425-1

Composition for this book was provided by the editors.
This book was produced without formal editing by the publisher.

Printed and bound in the United States of America

The paper used in this publication meets the requirements of the American National Standard for Permanence of Paper for Printed Library Materials Z39.48-1984.

6   5   4   3   2   1

Contents

List of Tables and Figures vii

1 Introduction 1

PART 1

CASE STUDIES OF INDUCED CHANGE IN AGRICULTURE POLICY

2 Economic Security in the Countryside: The Impact of Agrarian Change and Public Policy in Honduras, *Charles D. Brockett* 15

3 Brazilian Agriculture and the Debt Crisis, *Bruce Drury* 49

4 Agricultural Policy and Food Security in Peru and Ecuador, *Cynthia McClintock* 73

5 President Marcos, Multinationals, the World Bank, and the U.S. Government: Domestic and International Political Economy of Philippines' Coconut Industry, *Gary Hawes and Gretchen Casper* 131

6 Problems of Agricultural Development in Turkey, *Mahir Fisunoğlu and Birol A. Yeşilada* 151

7 The Food Crisis in Kenya, *Taye Woldesmiate and Ron Cox* 181

PART 2

INTERNATIONAL FACTORS AND AGRICULTURE IN THE THIRD WORLD

8 The Impact of Dependency on Agriculture and Food Crises in the Third World, *Birol A. Yeşilada* 205

9	Political Implications of International Monetary Fund Conditionality for Latin America, *Adalberto J. Pinelo*	239
10	Innovation in the IMF: The Cereal Imports Facility, *Valerie J. Assetto*	263
11	Who Governs the Rome Food Agencies? *Ross B. Talbot and A. Wayne Moyer*	281
12	U.S. Foreign Agricultural Policy and the Less Developed Countries, *Jean Doyle*	305
Contributors		337

Tables and Figures

TABLES

1.1	Energy deficient diets in developing countries	3
2.1	Daily food consumption calories and protein per capita, 1964-1966 and 1979-1981	17
2.2	Major commodity exports, 1960-1964 to 1980-1981	20
2.3	Agricultural land use: Changes in area devoted to export crops relative to food crops, 1948-1952 to 1981-1983	22
2.4	Changes in production, area and yield for major crops, 1962-1965 to 1981-1982	23
2.5	Change in food supply	24
2.6	Beef production, exports, and supply, 1961-1964 to 1980-1981	27
2.7	Change in the distribution of farmland, 1952-1974	29
2.8	Land distribution under the reform process ..	38
3.1	Value of major agricultural exports, 1973, 1981-1984	57
3.2	Production figures for major agricultural crops, 1970-1983	59
3.3	Changes in agricultural production of major crops, 1973-1983	60
3.4	Area planted in major crops in thousands of hectares, with 1973 to 1983	64

3.5	Yield of major crops in kg per hectare with 1973 to 1983	65
4.1	Trends in value of agricultural production: Ecuador	77
4.2	Trends in value of agricultural production: Peru	78
4.3	Trends in the value of agricultural production under democratic government	79
4.4	Production of food staples: Ecuador	81
4.5	Production of food staples: Peru	82
4.6	Landholding patterns in Ecuador	86
4.7	Agricultural yields: Ecuador and Peru	89
4.8	Changes in agricultural patterns: Peru and Ecuador	90
4.9	Trends in farmgate prices in Ecuador	98
4.10	Patterns of credit distribution by landholding size: Ecuador	99
4.11	Trends in farmgate prices in Peru	101
4.12	Public investments in agriculture: Ecuador and Peru	105
4.13	Peruvian and Ecuadorian irrigation projects: Relative costs	108
4.14	Trends in agricultural investment in Ecuador	110
4.15	Trends in agricultural investment in Peru	111
6.1	Agricultural support prices	158
6.2	Indicators of agricultural modernization	162
6.3	Agricultural production figures	165
6.4	Index of agricultural exports	165

6.5	Average annual change in daily wages of workers	168
6.6	Land distribution in Turkey	171
7.1	Agriculture and food production in Kenya	190
7.2	Balance of payments and external debt for Kenya	196
8.1	Current international commodity agreements in agriculture	211
8.2	The frequency of application of various non-tariff barriers in industrial countries, 1984	221
8.3	Current account balance as percentage of GNP in peripheral countries	223
8.4	Gross domestic savings and investment in peripheral countries	225
8.5	Debt indicators for peripheral countries	227
8.6	Distribution of debt reschedulings by region	230
10.1	Purchases in the CIF, 1981-1985	275

FIGURES

1.1	Food production per capita in developing countries	5
3.1	Changes in Brazilian agricultural production and population: 1973-1983	61
7.1	Index of agricultural production in Kenya	193
8.1	Composite index of commodity prices, 1950-1985	208
8.2	International commodity agreements, prices, and price ranges	212
8.3	Agricultural trade index by region	216

1

Introduction

Agrarian reform programs in the 1960s and then the Green Revolution were to have accelerated the modernization process so that hunger would be substantially eliminated throughout the world. Instead, the number of the malnourished in the Third World continues to grow. The cause of this tragedy is much more than the inexorable increase in population growth. At least equally important are patterns of land and food distribution and the underlying determinants of these patterns, namely, the interrelated factors of political power and economic structures.

Agrarian reform programs were much easier to formulate than to implement. In few countries did the configuration of political power allow meaningful redistribution of large inefficient holdings; rather, the usual result was their transformation to modern commercial operations along with the distribution through colonization of unused and less desirable lands. It was these commercial farmers, not the far more numerous subsistence peasants, who were most able to take advantage of the technological package of the Green Revolution. And, many of the peasants who were able to take advantage of the new techniques had to revert to traditional practices when the oil price shocks of 1973 and 1979 escalated the prices of fertilizers and fuels.

As commercial agriculture has spread throughout the Third World, it also has become increasingly integrated into international capitalist economy. A primary stimulus for many commercial growers has been the profit to be made producing for foreign markets. Their production has been facilitated by public policies of governments eager to expand their foreign exchange earnings. This motivation has been reinforced greatly by the debt crisis of the 1980s.

The abundant supply of capital created by the petroleum profits of the 1970s needed to be recycled by the international banking community. It found ready customers in Third World governments that wanted loans to cover costs of petroleum imports and development projects rather than cut back on their ambitious plans. These escalating foreign debts collided with the harsh world recession of the early 1980s, creating economic crisis throughout the Third World. Just to service their debts, much less repay them, now countries must further direct their economies toward available export markets, which for the Third World often means agricultural commodities.

If agrarian reform means progressive changes intended to create a more egalitarian rural society, then these recent dynamics have produced agrarian reform in reverse. Smallholders have been squeezed out of land markets and sometimes coerced off of their land. When combined with population growth, a rapidly increasing rural proletariat has been created, often unemployed or underemployed. Numerous others migrate to urban areas, often just transferring the problem. Land that previously had been planted in food staples now produces cotton, pineapples, sugar cane, soybeans, groundnuts, etc., or grazes cattle for export, usually to more developed countries. As domestic food production stagnates, it has fallen behind population growth rates in many countries.

Estimates of world hunger vary depending upon the method used but all agree that the number is huge. The Food and Agriculture Organization (FAO) claims that 436 million people suffer from undernourishment (FAO, 1985: 5) while the World Bank (1986: 3) put the number at 730 million, as shown in Table 1.1. The proportion of people with deficient diets in 1980 was virtually unchanged from 1970 and, therefore, the absolute number has increased. The World Bank (1986: 18) has estimated that the ten percent diet deficient category increased by ten percent and the twenty percent diet deficient category increased by fourteen percent. It has been claimed that in the 1980s, starvation and hunger-related diseases still cause 35,000 deaths per day and 13-18 million deaths per year (Hunger Project, 1985: 7).

The causes of this continuing tragedy have many dimensions. This volume concentrates on the political-economic. The contributors are organized into two sets. Part 1 includes case studies from each of the four regions of the Third World: Africa (Kenya), Asia (Philippines), Latin America (Brazil, Ecuador, Honduras, and Peru), and the Middle East (Turkey). There are certainly situations and

Table 1.1

Energy Deficient Diets in Developing Countries, 1980
(millions of individuals)

Country Group(n)	10% deficiency[a]		20% deficiency[b]	
	population	% of pop.	population	% of pop.
All LDCs (87)	730	34	340	16
Low-income (30)	590	51	270	23
Middle-income (57)	140	14	70	7
Sub-Sahara Africa (37)	150	44	90	25
E. Asia and Pacific (8)	40	14	20	7
South Asia (7)	470	50	200	21
Middle East and North Africa (11)	20	10	10	4
Latin America and Caribbean (24)	50	13	20	6

[a] 10 percent deficiency means that the person is receiving 90 percent or less of the FAO/WHO requirement and does not have enough colonies for an active working life.

[b] 20 percent deficiency means the person is receiving 80 percent or less of the FAO/WHO requirement and does not get enough calories to prevent stunted growth and serious health risks.

Source: World Bank (1986:17).

determinants unique to each region and country, as suggestively portrayed by Figure 1.1, for example, which presents the varying regional patterns of per capita food production in recent years. Nonetheless, each of these examinations of the relationship between agricultural policy and food security reinforce this volume's theme of "agrarian reform in reverse." The analysis shifts to the international level in Part 2. In this section the contributors examine various international factors that influence agricultural development in the Third World. These factors are: dependency in international trade and monetary relations, the IMF conditionality in Latin America, the Cereals Imports Facility of the IMF, the Rome Food Agencies, and the U.S. foreign agricultural policy toward the Third World.

PART 1: CASE STUDIES OF INDUCED CHANGE IN AGRICULTURE POLICY

The Latin American case studies are of very different countries yet they portray similar forces at work with similar consequences. Honduras is one of the poorest countries in the region and one of the most rural. Nonetheless, there has been substantial change in the Honduran countryside in recent decades, as the Brockett chapter demonstrates. The spread of modern commercial agriculture has advantaged part of the rural population but, when combined with rapid population growth, has diminished economic security for many others. Responding especially to the stimulus of export markets, commercial farmers have expanded their holdings and production. Conversely, subsistence farmers have lost access to land and the rural population to food. Across recent decades, per capita domestic food production has declined as has per capita beef supply; meanwhile, the relative share of land devoted to export crops as compared to food crops has increased substantially, as have beef exports. The second part of this chapter examines the political context of these socio-economic changes. Domestic and international political actors have promoted agricultural modernization and export expansion. At the same time, the state has had to contend with a peasantry that mobilized in response to the deleterious effect of these agrarian changes on their lives. The most important of the state responses has been an intermittent program of land reform, one that has been unable to keep pace with the growing numbers of landless and landpoor peasants.

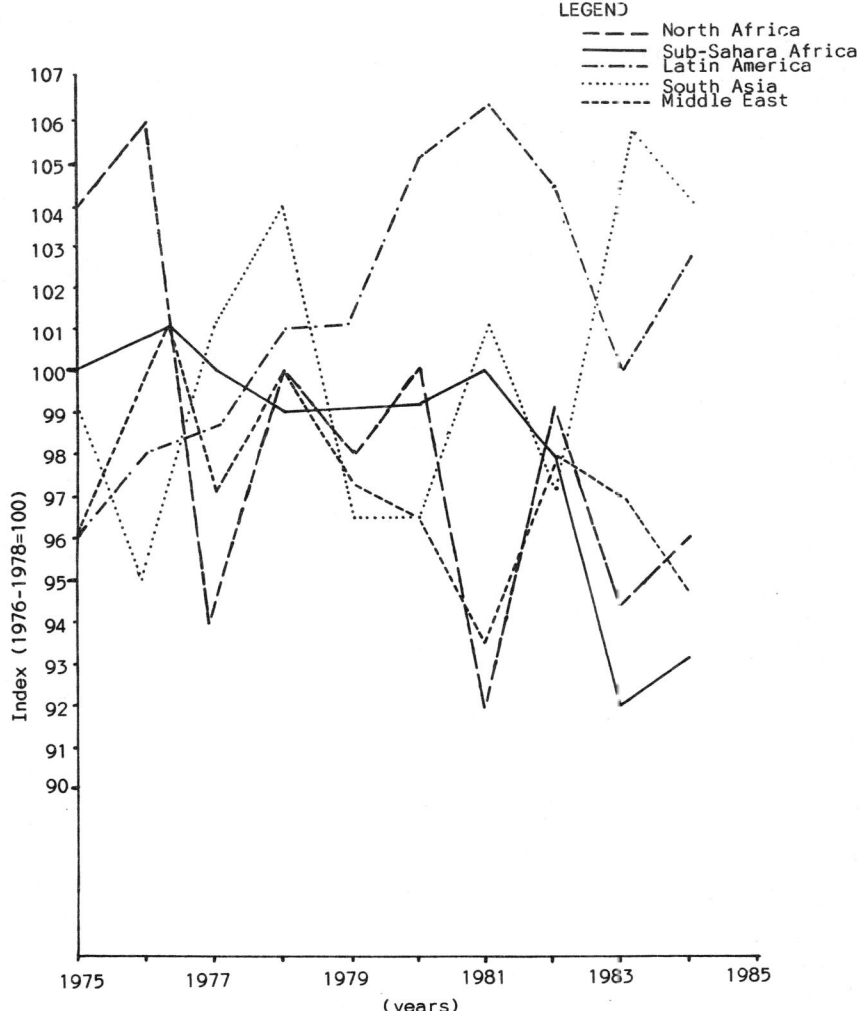

Figure 1.1 Food Production Per Capita in Developing Countries

Source: USDA. <u>World Indices of Agricultural and Food Production, 1975-1984</u>.

Brazil has a gross national product about eight times larger than that of Honduras, a population about 31 times larger, and has almost twice the percentage of its people living in urban areas. In their agricultural sectors, though, similar forces have been at work in recent decades. Drury's chapter documents the successful expansion of export agriculture in Brazil. At the same time, however, per capita production of major food crops stagnated or declined across the decade ending in the early 1980s. The two trends were related: land planted in the export crops rose dramatically while only minimally for the food crops. As consequences, caloric intake declined, inequality of land access increased, and rural unemployment rose. In explaining these dynamics, Drury gives particular attention to the debt crisis and the consequent need to increase foreign exchange earnings through agricultural export expansion. A further result was to end already feeble attempts at land reform. Agrarian reform in reverse in Brazil also has led to increasing rural tensions; conflicts over land provoked an average of twenty deaths a month in 1986 (Lanfur, 1987: 11). "Export agriculture," as Drury points out, "has defeated food and reform."

The McClintock chapter doubly enlarges the comparative focus of the case studies of Part 1. Her two countries of Ecuador and Peru fall between Brazil and Honduras in per capita GNP; furthermore, her case material allows a comparison of two major competing approaches to agrarian policy. On the one hand, a progressive military regime in Peru undertook a major land redistribution during 1968-1975, organizing much of the reform sector into cooperatives. In contrast, the promotion of capitalist agriculture with little attention to land reform has characterized agrarian policy in Ecuador throughout the period examined and in Peru under an elected civilian government during 1980-1985. Through her in depth analysis of public policy and its consequences, McClintock finds that although the Peruvian reform often has been faulted, the performance of its agricultural sector consistently equals or surpasses its Ecuadorian counterpart on the indicators studied. One notable difference in the two approaches concerns export agriculture. Public policy in Peru during the reform period attempted to redirect the traditional export commodities to the domestic market; in Ecuador, however, the contribution of export crops to the value of agricultural production more than tripled from 1970 to 1985 while production of food staples stagnated. At the more fundamental level of food security, though, what is most striking is the similarity in results: per capita food production has declined in both countries, necessitating increases in food imports that inadequately meet growing needs. The causes are familiar, ranging from the power of landed and other elites to resist or undermine reforms, government policies that favor larger landowners, world recession, foreign debt crisis, and pressures from the IMF and the U.S.---many of which McClintock summarizes as these countries' more thorough "integra[tion] into the world capitalist economy."

The next chapters expand the points of comparison, not only by bringing in the other regions of the Third World, but also by providing variations in population size and level of development. With populations around fifty million in the early 1980s, both the Philippines and Turkey have much larger populations than the other countries, with the exception of Brazil. In per capita GNP, Turkey falls between Brazil and Peru while the Philippines ranks between Peru and Honduras. Kenya has a population about the same size as Peru but with a per capita GNP almost one-third less, making it the poorest country among those examined in this volume.

The Hawes and Casper chapter on the Philippines provides a different approach to the analysis of agrarian reform. Their major concern is the role of the state in promoting the economic transition to export-led industrialization. As they point out, some predicted in the early 1970s that the Marcos regime would join the group of East Asian success stories. To some observers, a strong leader and strong state provided the social control and policy direction that would prove economically successful when combined with substantial support from international lenders. Instead, the Marcos regime brought economic crisis and its own demise. Hawes and Casper illuminate the causes of this tragedy through an examination of the coconut industry, the second most important industry in the Philippines and generator of about 25 percent of export earnings and of up to one-third of the population's income. What they demonstrate is that the Marcos regime concurrently centralized its control over coconut processing and marketing while expropriating an increasing amount of the surplus generated by this community. Rather than rationally utilize this control and surplus to further the development of both the coconut industry and the overall economy, it was instead squandered through massive corruption and regime legitimation efforts. As they conclude, the interests of the regime and of the state turned out not to be the same. While the other case studies raise serious issues about the probable consequences for the rural population of the original development project envisioned by other state actors and international lenders, that project was distorted by the self-serving policies of Marcos and his cronies. The Philippines of the Marcos period, then, is a good illustration of the theme of this volume as surplus was expropriated from coconut producers, including many small and medium producers, to further the interests of the dictatorship.

Fisunoğlu and Yeşilada's chapter on Turkey concentrates on a comparison of the policy impacts of different development models, in this case an import-substitution model from 1963 to 1980 and then an export-led growth model combined with an IMF austerity program since 1980. They find that the difference matters. During the import substitution period, public policy benefitted production of both food and export crops; per capita domestic food production increased across the period as did exports of traditional commodities. Domestic political rigidities during this period, though, prevented adequate responses to the oil shock of 1973 and changing economic needs. In crisis by the end of the decade, Turkey adopted an IMF austerity package and switched policy focus to export promotion. Since then, production of food crops has lagged behind population growth and wage increases behind prices. Yet, the foreign debt continues to grow, consuming more of export earnings, creating an even further imperative to accelerate export expansion. Fisunoğlu and Yeşilada also discuss land reform policy in Turkey since 1923. While plans have been formulated and minor programs implemented, landed interests and their urban allies have been able to frustrate major reform. As a consequence, they find that land concentration has increased and the number of landless rural families continues to grow.

The most recently decolonized of the countries included in this volume, Kenyan agriculture is still heavily influenced by the structures created by the British, as shown in the chapter by Woldesmiate and Cox. A dualistic agricultural system was created during the colonial period with a white elite controlling much of the best land, which was organized as large estates producing for export, with the indigenous population restricted to subsistence farming. With Independence, a Kenyan elite (and multinationals) replaced the colonial elite but the dualistic structure remained intact. In fact, they argue that the gap between the two has worsened as elite influence over public policy has insured a developmental model that promotes the interests of the export sector over the interests of smallholders. This approach has been reinforced by the growing debt crisis of recent years and the acceptance by Kenya of three agreements with the IMF. The results are the same as demonstrated by the other cases: production of export crops has grown much faster than that for food crops---indeed, per capita production of food crops has declined in some cases. With so much of the land devoted to large export estates, increasing numbers of Kenya's rapidly growing rural population are left with insufficient land to support a family or no land at all.

As the contributions of Part 1 demonstrate, domestic and international factors are interrelated causes of the continuing tragedy of hunger specifically, and poverty more generally, in Third World countries. Among the domestic factors discussed in these chapters, the following are the most important: market forces and public policies that advantage large producers; urban-biased development plans; and self-interest of governing officials. The most important of the international factors analyzed range from international market forces to the plans and pressures of actors such as the U.S. and the IMF to the actions of multinational corporations. So important are these international forces to understanding today's Third World agrarian systems that they are given direct attention in Part 2.

PART 2: INTERNATIONAL FACTORS AND AGRICULTURE IN THE THIRD WORLD

In an environment of globally expanding capitalism, agricultural development in Third World countries is no longer a simple domestic concern, affected solely by international forces. Increasingly, the agricultural policies of these countries are being influenced, if not determined, by international factors over which Third World states have almost no control. Such factors include the protectionist trade policies of the industrial West, artificially determined primary commodity prices, international monetary system, the West's domination of powerful institutions like the IMF and World Bank, problems in the Rome Food Agencies, and the foreign agriculture policies of the U.S. The contributors to Part 2 examine the implications of these international factors on agricultural development and food security in the Third World.

The Yeşilada chapter looks at the nature of the exchange relationship between developed and Third World states. He identifies several problems that hinder agricultural development in the Third World. These problems encompass falling commodity prices, protectionist agricultural trade policies of the industrial Western states, the international debt crisis, and the growing involvement of the IMF in domestic development policies of Third World states. Yesilada concludes that the nature of the present exchange relationship between the developed and developing countries can best be characterized by deepening dependency, The growing concerning of Third World states for the acquisition of foreign exchange to service their debts at all costs presents serious problems for agrarian reform in these countries. On the one hand, Third World states need to expand their export of agricultural goods to acquire hard currency to service their debts and to invest in domestic development projects. On the other hand, the industrial West's trade policies and the Third World states' push for the increased production of cash crops threatens the traditional food base of the developing world.

Pinelo continues the analysis of IMF austerity policies by an expanded look at Latin American states. In this chapter Pinelo argues that in recent years, the IMF represents not only a traditional actor with an expanding role in international monetary relations but also a new vehicle for the exercise of U.S. influence in Latin America. In this context, Pinelo examines the delicate problems the new Latin American democracies face as they attempt to meet the needs of their citizens while the IMF conditionality often demands contradictory decisions.

Assetto's chapter expands further the examination of the IMF by analyzing the Cereals Imports Facility (CIF). The IMF established the CIF in 1981, at the urging of the World Food Council and the Food and Agriculture Organization, to ameliorate the international food insecurity problem, especially in low-income developing countries. Assetto argues that the CIF, as it is presently constituted, suffers from several defects which serve to inhibit its use by low-income developing countries and, thus, reduce its efficacy. The linkage of quotas of the amount of CIF funds for which a member is eligible reduces the absolute amount of assistance upon which that member can draw in periods of emergency. This increased conditionality of CIF drawings also inhibits borrowing from the facility due to higher political and economic costs. Assetto concludes that unless the political conditions within the IMF and in the broader international distribution of power change significantly, the CIF will remain of limited value to the very states which need it most.

The analysis of international institutions is broadened in the Talbot and Moyer chapter where the authors examine the four world food organizations headquartered in Rome---the Food and Agriculture Organization (FAO), the World Food Council (WFC), the World Food Programme (WFP), and the International Fund for Agricultural Development (IFAD). They attempt to describe, explain and analyze the configurations of power which have prevailed within and between these organizations, and engage in a limited amount of forecasting as to what the future may bring. In their forecasting Talbot and Moyer consider unification of the food agencies, major expansion of their authority, dissolution, and incremental change within these organizations. The authors conclude that these institutions have functioned somewhat successfully, with some positive results and are necessary instruments to be used in the pursuit of justice and equity, both for rich as well as poor nations.

The final chapter by Doyle examines the impact of the U.S. foreign agricultural policy on less developed countries (LDCs). She traces the origins of the U.S. foreign agricultural policy, changes in this policy since the 1973 world food crisis, and the Reagan factor in recent years. Doyle argues that the content of the U.S. foreign agricultural policy is determined by the interplay of domestic and international, political and economic, practical and ideological factors, which are constantly reordered as conditions change. Furthermore, since WW II, this policy has been linked to other interests of the U.S. As a result of this linkage, U.S. foreign agricultural policy has served to extend and preserve U.S. global interests. It has provided the U.S. an entree into the LDCs, either directly through bilateral aid programs, or indirectly through its multinational corporations, control of transnational institutions, and participation in multilateral assistance efforts aimed at economic and agricultural development.

The contributors to this volume demonstrate that the provision of adequate food for all has become an increasingly critical problem in the Third World. In the 1960s and early 1970s, a combination of political and technological progress---agrarian reform and green revolution---seemed to provide hope for the elimination of hunger. However, the countervailing forces in the world economy in the 1970s and 1980s---soaring agricultural costs, the high cost of money needed to finance development projects, the destruction of traditional domestic markets, and the destabilization of world markets---have resulted in an increasingly uneven distribution of wealth and land and a new Third World food crisis.

REFERENCES

FAO (1985). *The State of Food and Agriculture, 1984*. Rome: FAO Publications.

Hunger Project (1985). *Ending Hunger: An Idea Whose Time Has Come*. New York: Preager.

Langfur, Hal (1987). "Brazil's Land Reform Program is Caught in a Violent Cross Fire." *Christian Science Monitor* (7 May).

U.S. Department of Agriculture (1984). *World Indices of Agricultural and Food Production, 1975-1984*. Washington, D.C.: USDA.

_____ (1985). *World Indices of Agricultural and Food Production, 1976-1985*. Washington, D.C.: USDA.

World Bank (1986). *Poverty and Hunger: Issues and Options for Food Security in Developing Countries*. Washington, D.C.: World Bank Publications.

PART 1

CASE STUDIES OF INDUCED CHANGE IN AGRICULTURE POLICY

2

Economic Security in the Countryside: The Impact of Agrarian Change and Public Policy in Honduras

Charles D. Brockett

Honduras is a very poor country. On standard measures of socio-economic development, Honduras typically ranks third from the bottom among Latin American nations. Not surprisingly, the greatest concentration of its poor are found in the rural areas. In the two and a half decades since socio-economic development became an object of Honduran public policy and United States policy toward Central America, there has been some improvement in the quality of life for part of the rural population, but less than planned. Furthermore, significant elements of the peasantry have experienced a deterioration of their position, in part because of a high rate of population increase, but also, in part, because of the commercialization of agriculture.

Since the end of World War II some landowners have expanded and intensified their production, especially for exports, in response to market incentives and governmental encouragement. While agricultural modernization and export expansion often are justified as fundamental to the developmental process, in Honduras they have resulted in not only increased export earnings, but also declining access to land, income, and food in rural areas. Reinforced by population growth, rural insecurity has in turn lead to peasant organization, land invasions and occupations, and an agrarian reform program.

The object of this study is to assess the impact of public policy---both Honduran and that of the U.S.---on economic security in the countryside. The paper begins by documenting the extent of the expansion of export agriculture and the impact of this expansion on rural economic security---that is, access to a land, income, and food. The paper then examines public policy concerning export agriculture and land redistribution and other aspects of rural development programs.

ECONOMIC SECURITY

In 1984 Honduras had the lowest per capita gross national product ($660) and the highest population growth rate (3.4 percent) in Central America, including the countries of Belize and Panama. Among the seven nations of the region it has the worst ranking for infant mortality (76 infant deaths per 1000 births) and life expectancy (60.4 years) and the second worst for literacy (57 percent, Guatemala lower) and people per doctor (3,123, El Salvador higher). It is also one of the most rural countries in the region (61 percent live in rural areas) with the highest percentage of its work force employed in agriculture (60 percent, US/AID, 1985; 6). Typically, only Haiti and Bolivia rank lower than Honduras among the Latin American nations on such measures of socio-economic development.

While these indicators of development are low, they used to be even lower. For approximately the last two and a half decades it has been a stated goal of both Honduran and United States policy to improve the welfare of the Honduran people, both through the extension of services and through economic development. The improvement, however, has not been substantial.

A striking manifestation of Honduran poverty is the data on malnutrition. Table 2.1 shows that the daily per capita consumption of calories and proteins has increased over the last two decades (by nine and 2.2 percent, respectively) but not at all since 1969-1971 for calories and with a three percent decline for proteins. Furthermore, average daily caloric intake in 1979-1981 was still six percent below the FAO/WHO recommendation for Hondurans. Since this is an average figure, then clearly many consume less than this, some far less. One study has estimated that in 1973, 60 percent of the Honduran population consumed at least ten percent less than the daily recommended caloric requirement (Reutlinger and Alderman, 1980: Appendix Table 2A).[1] The situation is worst for people in rural areas, especially children. In 1978 it was estimated that 64 percent of the rural population was unable to meet minimum dietary needs; for rural children under the age of five the corresponding figure was 90 percent (Morris, 1984: 22,24). Studies in the early 1970s found the third highest rate of moderate and severe malnutrition in the hemisphere among children under

Table 2.1
Daily food consumption calories and protein per capita, 1964-1966 to 1979-1981

	1964-1966	1969-1971	1974-1976	1979-1981	% Change
Honduras					
calories	1,959	2,132	2,074	2,135	9.0
proteins	51.0	53.8	50.5	52.1	2.2
Central America					
calories	2,127	2,128	2,242	2,254	6.0
proteins	57.1	58.8	57.8	57.7	1.1
United States					
calories	3,341	3,497	3,513	3,641	9.0
proteins	102.8	105.7	105.8	105.6	2.7

Source: FAO. *Production Yearbook, 1983* (1984)

the age of five in Honduras, after Haiti and Guatemala. Other studies suggest that the prevalence of malnutrition among such children increased by 80 percent in Honduras from the mid-1960s to the mid-1970s, paralleling trends throughout Central America (IDB, 1978: 138,141).

Clearly, the rapid growth of the Honduran population is a primary cause of these manifestations of Honduran poverty. The descriptive literature on Honduras and empirical studies of other Central American countries, however, suggest other forces at work as well. In his landmark study, Durham (1979:21-51) demonstrated that land scarcity in El Salvador has been the result not so much of population growth as it has been the consequence of increasing land concentration that occurred with the expansion of the production of crops for the export market. At the same time that per capita food production declined during 1950-1970, per capita land devoted to export crops actually increased. Looking at the same dynamics in Guatemala, I found the same results (Brockett, 1984: 480-487). From 1948-1952 to 1980-1981, the production per capita of basic food crops declined seven percent while the relative share of land devoted to export crops increased ninety percent compared to the share devoted to basic food crops.[2]

While the evidence has yet to be systematically evaluated, several authors have described similar changes in contemporary Honduras. Although Honduras has had a long experience with export agriculture, it had relatively little impact on rural society until the post-1950 period (Durham, 1979; Parsons, 1976; Pfeil, 1977; Ruhl, 1984; Volk, 1981). The traditional export crops of Central America have been coffee and bananas. The coffee boom of the late 1800s had substantial impact on both rural and national class structures where it occurred (primarily Guatemala, El Salvador, and Costa Rica) but it largely by-passed Honduras.
The rugged mountainous terrain of Honduras is often too steep for successful agriculture (60.8 percent of its surface area slopes at more than 40 percent), often lacks the fertility found in the rest of the region (it lacks its neighbors' volcanic ash), and usually impedes transportation between the highlands and coastal ports (the capital, in the highlands, still lacked an all-weather paved road to the north coast until 1970). Bananas, on the other hand, have dominated the economy since the beginning of this century.
Not only are they the country's leading export, peaking at 88 percent of exports during 1925-1939, but throughout this century the U.S. companies that control the banana industry have played a leading role in politics, from the financing

of revolutions in the early years to bribing a government (and indirectly toppling it) in the "Bananagate" scandal of the 1970s (LeFeber, 1984: 44-46, 207-208). Bananas in Honduras, however, are an enclave economy. Grown on the north coast, the banana region has been only loosely connected to the rest of the country, at least until 1954.[3] While foreign banana companies obtained concessions for over one million acres of land by World War I, traditional land holding patterns were little affected since malaria, remoteness, and plentiful land elsewhere had kept the north coast relatively unpopulated until the advent of the banana companies.[4]

Except for the export banana industry on the north coast, until after World War II agriculture in Honduras was of the traditional Latin American style. Most land holdings were worked by families living at the subsistence level. Much of the land, however, was held in larger units which used the land inefficiently, leaving it fallow, grazing it with cattle, or employing share croppers. Honduras was unique in one respect, though, and that was the large amount of land which was outside of private ownership. As late as 1965-1966, 12.5 percent of the land in farms was nationally owned and another 21 percent was *ejidal* (community owned) land (Fonck, 1972: 30).

Market incentives and government encouragement after 1950 had a profound effect on this traditional rural society. Following World War II, escalating cotton prices prompted larger farmers throughout Central America to convert land to this new export crop. Within Honduras, the construction of the Pan American Highway through the southern part of the country, along with the construction of connecting highways to Tegucigalpa in the interior and San Pedro Sula in the north coast region, more effectively integrated the country, broadening the internal market and better linking agricultural producers to international markets. Beginning at about the same time, the government of Honduras began to encourage the diversification of commercial agricultural production as part of plans for economic development and as necessary for the expansion and diversification of export earnings.

In response to such incentives and encouragements, land planted in coffee, cotton, and sugar and land devoted to cattle raising greatly expanded. Export earnings have been expanded and diversified, as demonstrated in Table 2.2. While the five major agricultural commodities constituted two-thirds of total commodity exports in both the early 1960s and early 1980s, the relative shares of the five have

Table 2.2
Major commodity exports, 1960-1964 to 1980-1981

Product	1960-1964	1965-1969[a]	1970-1974	1975-1979	1980-1981
Bananas	45%	47%	38%	24%	28%
Coffee	16	13	15	28	25
Beef	3	3	7	6	7
Cotton	2	3	1	2	2
Sugar	--	--	1	1	4
SUBTOTAL	66%	66%	62%	61%	66%
Wood	11	9	14	8	3
Minerals	6	5	7	7	6
Tobacco	1	1	1	2	2
TOTAL	84%	81%	84%	78%	77%

[a]1968-1969 for wood, minerals, and tobacco.

Sources: First five commodities through 1975-1979, Weeks (1975: 77); bottom three and 1980-1981 for all commodities, United Nations, Yearbook of International Trade Statistics (various years).

significantly altered. Bananas, once 88 percent of export earnings, had fallen to 45 percent in 1960-1964 and down to 28 percent in 1980-1981. On the other hand, over the last two decades the share of coffee has increased from 16 percent to 25 percent, beef from 3 to 7 percent, and sugar from 0 to 4 percent. Cotton has been less successful. Its share of export earnings reached three percent in the 1965-1969 period but then fell, due to falling world prices, insect damage, and the disruption of the 1969 war with El Salvador (Blutstein, 1971: 147), although lately it has partially recovered.

While the expansion of commercial export agriculture has been beneficial in some respects, it has been claimed by several scholars that in Honduras it also lead to the conversion of land from food to export production, land concentration and landlessness, and rural unemployment and underemployment. The following sections evaluate the relevant evidence from the recent Honduran experience.

Export Crops and Food Production

Clearly there has been a tremendous expansion in the amount of land in Honduras devoted to the production of export crops. As Table 2.3 indicates, over the three decades ending in 1982 export land almost doubled. Meanwhile, the amount of land devoted to the production of Honduras' four major food crops increased by only 22 percent. Consequently, the relative share of land allocated to export crops increased from 36 percent of food crops around 1950 to 55 percent in the early 1980s---an increase of more than 50 percent. Registering the most notable increases were land in coffee (especially up to the early 1970s) and in sugar (especially since the mid-1970s).

It should be noted that the distinction between the two types of crops is not as clean as this table suggests. Honduras did not become self-sufficient in sugar until 1962 nor export sugar until 1967. Furthermore, domestic demand increased rapidly (Blutstein, 1971: 147) and as late as 1972, 75 percent of production went to the domestic market (World Bank, 1978: Annex 12, p. 9). Cottonseed oil is also the source of cooking oil and margarine. On the other hand, large quantities of both corn and sorghum are used as animal feed---28 and 48 percent respectively in 1981 (USDA/FA, 1981)---and therefore are indirectly related to the export market. It also should be pointed out that the land held by banana companies is much greater than that reported here.

Table 2.3
Agricultural land use: Changes in area devoted to export crops relative to food crops, 1948-1952 to 1981-1983

Crop	1948-1952	1961-1965	1971-1973	1974-1976	1981-1983	% Change
Beans	50[a]	73	71	70	74	48
Corn	283	275	297	333	337	19
Rice	11	9	14	18	23	109
Sorghum	57	42	31	53	56	-2
TOTAL FOOD	401	399	413	474	490	22
Bananas	57	42	50	50	50[b]	-12
Coffee	63	83	107	110	121	92
Cotton	1	7	5	7	7	600
Sugar	22	33	53	49	93	323
TOTAL EXPORT	143	165	215	216	271	90
EXPORT as % FOOD	36	41	52	46	55	53

[a] figures are for area harvested in 1000 hectares. Many are estimates by either FAO or country.
[b] data not available; figures from latest year available.

Source: FAO. Production Yearbook (various years).

It has been their traditional (and controversial) practice to hold reserves substantially greater than the land actually planted (the figure reported in this table).

Area devoted to the four basic food crops has increased, as has yields, and therefore so has production (see Table 2.4). Unfortunately, though, this increase in production has not kept pace with population increase, as the following table documents. While Table 2.5 shows that total production of the four basic food crops has increased by 115 percent in the three decades since about 1950, the population of Honduras had increased by 183 percent to 1983. Consequently, the per capita production of the basic food crops actually declined across the three decades by 21 percent.[5]

Table 2.4
Changes in Production, Area and Yield for Major Crops, 1962-1965 to 1981-1982

	Production	Area	Yield
Food Crops			
beans	-6%	4%	-10%
corn	66	24	35
rice	236	144	2
sorghum	14	40	23
TOTAL FOOD	49%	25%	13%
Export Crops			
bananas	50	19	88
coffee	159	43	137
cotton	55	0	66
sugar	374	179	31
TOTAL EXPORTS	185%	62%	53%

Source: FAO. Production Yearbook, 1970, 1983 (1971, 1984)

Table 2.5
Change in food supply[a], 1948-1952 to 1981-1983

Crop	1948-1952	1952-1956	1961-1965	1971-1973	1974-1976	1981-1983	% Change
beans	22	22	48	47	32	45	105
corn	205	204	298	331	363	487	138
rice	18	18	11	15	28	39	117
sorghum	47	48	51	43	49	56	19
Total Production	292	292	408	436	472	627	115
Production per capita[b]	203	180	200	159	154	160	-21

[a] domestic production of four major food crops reported in metric tons. Many estimates, by either country or FAO.
[b] kilograms per capita.

Source: FAO. Production Yearbook (various years).

Until the 1970s, Honduras was an exporter of corn and beans, in fact, the leading exporter in Central America of those commodities (Fonck, 1972: 12). While it would be a mistake to describe it as self-sufficient in those commodities (as some writers did), since most of the population was malnourished, such exports did contribute to foreign exchange earnings. Since then the situation has often been reversed. Not only have wheat imports been increasing (Honduras produces only a minor amount of its wheat) but in several recent years since the mid-1970s more corn has been imported than exported. When imports of these five foods crops are taken into account, then the total supply per capita increases one percent from 1976 to 1981. It should be remembered, though, that imported food is not fully comparable to that which is domestically produced. What a rural family would be able to produce for itself if it had access to land, it might not have the income to purchase in the market place.

This decline in per capita production of the major food crops is directly related to the expansion of export crops in at least four ways. First, land was converted from food to export production. There are a number of descriptive accounts of this process. Indeed, some have referred to it as an "enclosure movement" as large landowners pushed subsistence farmers off of desired land through coercion (Parsons, 1976: 11-15). Sometimes the coercion was monetary; Parsons cites rents altered from sacks of grain to cash, as well as the rapid jacking up of cash prices (p. 15). Sometimes the coercion was physical; Durham (1979: 122) gives the example of another study which describes two haciendas which added 22,000 hectares to their holdings in this fashion. Second, new land was opened up for cultivation---30 percent more during the years covered in Table 2.5. Much of this land could have been devoted to the production of food for the growing population; instead, 65 percent went to the four export crops. Third, peasants desiring to buy their own land have had to face not just decreasing supply but also its inevitable companion, higher prices. They faced not only the competition of other farmers, large and small, but also urban business and professional people who were attracted to agriculture by the new profit-making possibilities (Parsons, 1976: 14). Finally, mention should be made of the vast holdings of the banana companies. It is true that they were obtained earlier in the century largely without disturbing landholding patterns. Yet, as land started to become scarce in the last several decades, their holdings were "plainly needed by, but inaccessible to, a rural population that had quadrupled since 1900" (Durham, 1979: 117). The banana company holdings in 1971 were estimated to still be around 200,000 hectares (one hectare = 2.47 acres); that is, about one-half of the total area planted at that time in the four basic food crops.

Paralleling these trends for food are those for beef production and export. The two are linked; the descriptive accounts almost always identify the desire to fence off new lands for cattle raising as a primary motive for the Honduran "enclosure movement." Cattle raising traditionally has been a primary activity in rural Honduras, but the nature of this endeavor has changed in recent decades. At one time about 80 percent of the cattle were raised by farmers whose herds averaged less than fifty heads. The relatively unimportant exports were often live animals. Cattle were usually sold by the head rather than by weight, thereby minimizing incentive for quality improvement (Blutstein, 1971: 149; Bakken, 1965: 155).

Expanding markets for beef domestically, and especially internationally, have significantly changed the cattle industry throughout Central America (Brockett, 1985; Shane, 1980). Since 1969, the majority of the cattle extracted in Honduras have been for the export market and virtually all have been slaughtered first (Slutzki, 1979: 113). As demand increased and sales came to be based on slaughtered weight, ranchers had the incentive to expand their holdings and to improve their grazing practices.

The results of these trends are indicated in Table 2.6, which shows that Honduras has replicated the Central American pattern to the extreme. While beef production (as measured by weight) has more than doubled over the last two decades, exports have expanded much more rapidly, expanding from 16 percent of production in the early 1960s to 51 percent of production by the late 1970s. Consequently, beef has become an important earner of foreign exchange; its contribution to export earnings increased from three to seven percent across the two decades to the early 1980s, moving it into third place among commodity exports (Table 2.2). By the late 1970s, Honduras had become the fourth leading exporter of beef to the U.S. among Latin American nations (Shane, 1980: 77). With both population and beef exports expanding rapidly, the per capita supply for domestic consumption, consequently, has declined. As measured by the evidence presented here, this decline has been 30 percent for the two decades up to the early 1980s.[6]

Ironically, throughout much of the late 1970s Honduras had a meat "surplus" (USDA/FAS, 1980). Access to the U.S.

Table 2.6
Beef production, exports, and supply, 1961-1964 to 1980-1981

	1961-1964	1969-1971	1975-1977	1978-1979	1980-1981	% Change
Honduras						
production	24	33	45.3	44.5	51	113
exports	3.6	12.9	18.4	22.6	26.2	563
exports/prod.	16%	39%	41%	51%	51%	219
prod. per cap.	11.2	12.9	14.4	12.5	13.5	21
supply per cap.	9.4	7.9	8.6	6.2	6.6	-30
Central America						
exports/prod.	15%	32%	34%	37%	29%	93
prod. per cap.	12.4	13.9	15.3	16.2	14	13
supply per cap.	10.6	9.4	10.1	10.2	9.9	-7

Sources and Units of Measurement:
Production: FAO, Production Yearbook (various years): metric tons of beef and buffalo meat from indigenous animals--some are estimates by country or FAO.
Exports: United Nations, Yearbook of International Trade Statistics (various years): metric tons of bovine meat, fresh frozen also, for Guatemala and Costa Rica, meat tinned n.e.s. or prepared.
Exports/prod.: exports as a percent of production.
Production and supply per capita: kilograms. Supply per capita is based on production minus exports.

market is restricted by a quota system which was established in response to pressures from the U.S. cattle industry (Shane, 1980:101-104). The Honduran domestic price is controlled and largely determined by external demand (World Bank, 1978: Annex 12, Appendix 1, p. 6). The U.S. quota is invariably filled, but demand in Honduras is often insufficient at prevailing prices to absorb the rest.[7] In other words, not only was land converted from domestic food production to cattle raising, a controversial practice in itself, but this conversion went beyond what was necessary to meet market demands.[8] A final, and longlasting, consequence of this conversion to cattle raising is the rapid exhaustion of soil fertility where tropical forests have been cleared (Shane, 1980: 21-27).

Economic Security in the Countryside

For many farmers the expansion of export-cropping has meant an increase in their income and their standard of living; this includes many small and medium size farmers who grow much of the coffee crop (Blutstein, 1970: 145; Molina Chocano, 1983). On the other hand, the expansion of export-cropping for many other rural people has limited their access to land, employment, and income, as well as to food, as documented above.

The ownership of land in Honduras is highly concentrated, although not as badly as in the neighboring countries of El Salvador and Guatemala.[9] As measured by Honduran agricultural censuses, the structure of land concentration changed little from 1952 to 1974 (see Table 2.7). In 1952, five percent of the farms held 57 percent of the land; in 1974, four percent held 56 percent of the land. Conversely, 16 percent of the land was held by the 75 percent smallest farmers in 1952 and by 79 percent in 1974. The most significant change is an increase in the smallest farm category (under one hectare) from 10 percent of all farms to 17 percent in 1974.[10] Alongside this increase in the prevalence of the _microfinca_ has been a rise in landlessness. After discounting the (relatively) highly paid workers in the banana industry, Ruhl (1984: 49) estimates the percentage of landless among rural families to have increased from 10 in 1952 to 32 in 1974. Consequently, a majority of rural families (51 percent) are classified as landless or land poor.

Small and large farmers differ not only in the amount of land they have but also in the security of their tenancy.

Table 2.7
Change in the Distribution of Farmland, 1952-1974

Farm size in hectares	1952 % farms	1952 % area	1974 % farms	1974 % area
Under 1	10	--	17	1
1 - 5	47	8	47	8
5 - 10	18	8	15	8
10 - 50	21	27	18	28
50 - 100	3	11	2	12
100 - 500	2	18	2	22
Over 500	.3	28	.2	22

Source: Calculated from Mark J. Ruhl, "Agrarian Structure and Political Stability in Honduras," Journal of International Studies and World Affairs, 21, 1 (February): 50.

The 1965-1966 agricultural census found that the probability of land ownership steadily increased with farm size; only 14 percent of the smallest farmers (under 1.4 hectares) owned their land while 94 percent of the largest (over 609 hectares) were owners. Ownership by a majority was not reached until the over 34.9 hectare category. On the other hand, 44 percent of the farms under 2.4 hectares were rented (47 percent of all farms were in this category) while less than 10 percent of the farms over this size were rented. Rental agreements were usually verbal and for less than a year (Fonck, 1972: 30-31). More recently, AID estimates that only one percent of farmers have fee simple property titles while three-quarters of farmers have insecure tenancy (US/AID, 1982: 1). As commercial agriculture has spread throughout the countryside, renters have found themselves vulnerable and often dispossessed of the farms they had operated. But, they have not been alone. As previously indicated, Honduras is unique in the preservation of its public lands---ejidal and otherwise. As late as 1974, they still constituted 33 percent of all farms (Ruhl, 1984: 45). There often have been disputes, however, about who enjoys legitimate access to such lands. Especially in the first years of commercialization, small farmers found themselves pushed off of public lands (Parsons, 1976: 15; Fonck, 1972: 30). Peasant resistance stiffened, though, and peasant groups scored a number of surprising successes in defending their claims, as will be discussed below.

While small farmers tend to intensively utilize their land,[11] large farmers---who control most of the land---tend to underutilize it, even with the spread of commercialization. In 1952 farmers with less than 10 hectares cultivated about 56 of their land, while farmers with over 50 hectares cultivated only seven percent of theirs (Durham, 1969: 127); those with four hectares or less had about 74 percent under cultivation (Honduras, 1954: 38). More recent data demonstrates the same patterns. A World Bank report on proposed irrigation projects gives data on land distribution and land use patterns for areas under consideration. Most of the land is in holdings over 10 hectares in size. Land in pasture for the project areas ranges from 41 to 71 percent with a 56 percent average,[12] while land devoted to corn and beans ranges from seven to 46 with a 25 percent average. It should be noted that these are usually good agricultural lands---good soil and certainly more level than most of the country (World Bank, 1978: Annex 8).

In summary, agrarian change and population growth have heightened economic insecurity for many rural Hondurans over the last three decades. In some cases government policy has contributed to these trends. In other cases government action has attempted to improve economic security in the countryside, but with varying degrees of success. A consideration of the purposes and impact of Honduran and U.S. rural development policy is the subject of the following section.

IMPACT OF PUBLIC POLICY

Until after World War II, the Honduran government had played virtually no role in the economy with the exception of its various dealings with the U.S. banana companies which lead to their eventual take-over of the country's north coast and export sector (Kepner and Soothill, 1967). As the government's commitment to economic development and then social development slowly and intermittently increased through the 1950s and 1960s so did government capacity, but as late as 1971 it was incapable of effectively utilizing and absorbing existing loans by international donors (Blutstein, 1971: 177; US/GAO, 1970: 4). In a communication to Congress at the end of the 1970s, an AID official assessed the commitment and capability of the Honduran government as follows:

> Since 1972, but particularly since 1975, the Government of Honduras has demonstrated an increased commitment and will to accelerate development, allocating necessary resources more rationally, including increasing markedly its efforts to reach more of the disadvantaged, particularly the rural poor. This commitment (sic) per-se comes 5-15 years after similar decisions in other Central American countries. Given these circumstances, AID is only recently able to work with the Government of Honduras to structure a wide spectrum of programs... (US/House, 1979: 245).

Major areas in which programs of the two governments have affected rural families include agricultural development and agrarian reform.

Agricultural Development

A study mission to Honduras in the 1950s characterized the prevailing level of agricultural development as being "as primitive as can be comprehended within the meaning of the term 'agriculture'" (Checchi, 1959: 52). Responsibilities for promoting the country's agricultural development were given to a Ministry of Agriculture established in 1951 and a National Development Bank established the year before.[13] Serious agricultural planning, though, is said not to have begun until the late 1960s (Fonck, 1972: 24) and recent reports still find the lack of trained personnel and institutional underdevelopment to be major barriers to successful government promotion of agricultural development (World Bank, 1978: Annex 11; US/AID, 1985: 127).

Nonetheless, government programs (usually financed by loans and credits from international donors) have had an impact. There have been programs from the beginning aimed at improving basic grains production, some with success (Checchi, 1959: 62), and others with failure (Tendler, 1976: 5). On the other hand, many programs have promoted export agriculture, whether intentionally or not. Such support has been especially important to the expansion of cotton growing (Checchi, 1959: 62) and cattle raising, the latter of which is best documented.

It has been estimated that during the two decades prior to 1980 over half of the loans of the World Bank and Inter-American Development Bank for agriculture and rural development in Central America promoted the production of beef for export (Keene, 1980:2). Honduran ranchers have benefited from World Bank loans of $6.6 million in 1974 and $34 million in 1976 and IDB loans in 1971 of $2.8 million, 1972 of $8.9 million, 1974 of $4.4 million and 1978 of $4 million---a total of $60.7 million from the two sources in the 1970s (Shane, 1980: 36, 42). International actors also have been critical in the development of the Honduran beef industry in another way. The packing plants licensed for export are largely controlled by foreign capital (not all U.S.), including the largest two which do about three-quarters of the exporting. It was this foreign capital which initiated the Honduran beef export industry (Slutzki, 1979: 166-169).

In recent years, given the U.S. Congressional mandate to direct programs to the "poor majority," AID has reduced its direct support for cattle development (Shane, 1980: 52). The Honduran government also has redirected its attention; in 1978, the government announced a $10 million cattle development project, utilizing international funds, aimed at assisting small ranchers (USDA/FAS, 1978: 2-3). Given the marketing problems discussed earlier, though, cattle development programs remain controversial. U.S. funding began in 1983 when AID initiated a seven-year Small Farmer Livestock Improvement project with a $10 million loan and $3 million grant for credit and technical assistance. The intended beneficiaries are farmers with holdings between 20 and 50 hectares and agrarian reform cooperative farms (US/AID: 1983). In justifying the project, AID (1983: 6-7) notes that the quality of the land received by the reform sector "was most often land unsuited for intensive agricultural purposes." While it would be suitable for livestock grazing, on the other hand, rich agricultural lands held by large absentee landowners are often used instead for livestock grazing.

Until recently, governments and international donors, often unself-consciously, promoted export agriculture and the interests of large commercial farmers. Since the 1970s, a concern for the plight of the rural poor (who, it was realized, were not receiving much through trickle-down approaches) and, more recently, a concern for food security have complicated rural development strategies (Nesman, 1981; Staatz and Eicher, 1984). The implementation of these new concerns can be most problematic. In her review of selected

AID programs in Honduras, for example, Tendler (1976: 28) concludes:

> Though AID has often required that credit institutions lend more for certain groups, certain crops, or certain types of activities, it has not been successful at getting credit institutions to lend much differently than they normally do.

More specifically, despite AID requirements that the National Development Bank alter its lending patterns and give more for basic grains and to small farmers and less for export crops and to large farmers, she found that the development bank failed to do so. The essential problem is that such requirements are contrary to established practices of the bank, as well as its perceived interests and associations. Credit to larger farmers for export crops is a well-established and successful business practice. Banks have no compelling interests of their own to change such practices, but might make promises to do so in order to obtain scarce capital. The commitment to abide by these requirements, however, was ambiguous at best. Furthermore, as Tendler points out, a "contempt for the peasant pervades the middle class and urban employees of service institutions" (p.32).

As a manifestation of the Reagan administration's economic policies toward Central America, in the early 1980s AID became more vigorous in its effort to encourage the Honduran government to foster export agriculture. The AID perspective at this time was that attempts by the government of Honduras:

> to promote exports can be characterized as sporadic and relatively unsuccessful. The government has not tried particularly hard to push for increased exports or to attract investors for export operations. Long delays in obtaining government approvals, especially where exploitation of natural resources is involved, have discouraged all but the most determined exporters (US/AID, 1984: 7).

As a result of "policy dialogue" between the two governments and AID assistance, in 1983 the Honduran government passed an Export Incentives Law, one manifestation of it becoming "more committed to altering its policies to favor exports." AID notes that a "significant involvement by U.S. businessmen in Honduras" export development effort is essential if the program is to be successful (pp. 9-10).

Meanwhile in 1984, the U.S. initiated a three-year $7 million Export Promotion and Services project which similarly is aimed at boosting non-traditional, including agricultural, exports by "Honduran entrepreneurs." Intended to provide both technical assistance and foreign exchange, the project is characterized as "a massive technology transfer" utilizing "the accumulated experience of highly experienced trade executives" (p. 42). The cost of these U.S. businessmen is budgeted at about $100,000 per year per advisor. In the same year AID also established a Honduran Agricultural Research Foundation. Planned to run eight years at a cost of $20 million, this project will provide applied research activities for both export and foodcrops (US/AID, 1985: 115, 122-123).

United States assistance programs with the objective of aiding export agriculture are now explicitly justified for the assistance they will give the landless and small farmers (US/AID,1985: 115). Even programs targeted to help agribusiness are justified by their "strong backward linkages with important employment and income effects on medium and small farmers" (p.176). While some of these farmers certainly will benefit, the earlier analysis in this paper demonstrated than the expansion of export agriculture has undermined the economic security of many others in the countryside, just as it has elsewhere in Central America under conditions of gross inequalities in the distribution of power, land, and other resources. Unlike their neighbors, however, Honduran peasants have been more successful in organizing, asserting, and defending their interests. Over the years they have staged successful land invasions, become a significant force in national politics, and have gained land redistribution and titling programs.

Agrarian Reform

Agrarian reform first came to Honduras in the early 1960s, largely in response to the external stimuli of the Cuban Revolution and the Alliance for Progress (Fonck, 1972: 28-29). The most liberal president to that point, Ramon Villeda Morales, established the National Agrarian Institute (INA) in 1961 and in the following year passed through Congress an agrarian reform law which was particularly aimed at the fruit companies' uncultivated land. While Villeda was one of the more popular Latin leaderswith the Kennedy administration, his agrarian reform law was most unpopular with the U.S. ambassador and leading U.S. legislators, not to mention the fruit companies. Ambassador Charles Burrows recalls that he told the Honduran president, "This is not a good law. It's going to cause you all kinds of trouble, and I think you ought to take a very, very close look at it" (Burrows, 1969: 14). Meanwhile, U.S. Senators put pressure on the administration to protect the interests of the fruit companies and to uphold the principle of adequate compensation, which meant, in the words of Wayne Morse, compensation in "hard cold American dollars" (U.S. Congress, 1962: 21614-21620).

While Standard Fruit indicated that it could live with the law, United Fruit had no intention of doing so. It stopped its planting program, throwing many Hondurans out of work. Coincidentally, Villeda had a trip to the United States scheduled. Prior to returning to Honduras he met with United Fruit officials with whom he had a "very satisfactory conversation." He promised them a revised law which, in Burrow's words, would be "livable for private interests" (Burrows, 1969: 16-17). Prior to Villeda's trip, Burrows had written to the State Department the following message:

> I am sure that The Fruit Company is in an excellent position and can probably get much of what it wants from the Honduran Government in terms of agrarian law revision, replacement of INA personnel or anything else. Please pass this on where it will do the most good (Burrows, 1962: 2).

The revision was delivered. On October 3, 1963, Villeda was overthrown by a military coup for reasons apparently unrelated to the agrarian reform (Burrows, 1969: 35-36; but, see Shaw, 1979: 139).

For the next decade any land redistribution was initiated by the peasants themselves. In response to the "enclosure movement" and land pressures discussed above, as well as to the betrayal of the promise inherent in the 1962 law, Honduran peasants organized and asserted themselves to a degree unparalleled in Central America to that point (Anderson, 1981: 159-162; Morris, 1984: 45-50, 79-81, 96-100; Parsons, 1976: 8-11; Pearson, 1980: 297-320; Pfeil, 1977: 77-144; Ruhl, 1984: 49-56; Volk, 1981: 14-23). By the late 1960s, the major peasant organizations claimed a combined membership of 90,000 rural families (Ruhl, 1984: 51). While any land redistribution depended on their initiative, under the leadership of Rigoberto Sandoval Corea as the head of the INA during 1967-1972, government support at times was forthcoming in their behalf. When peasant land occupations and invasions were contested by other interests (e.g., largeland owners), INA adjudicated the conflict, often siding with the peasants during this period.[14]

With the uncontested election of a conservative president in June, 1971, however, leadership and policy of the INA were soon to change. In response, peasants became more restive and on December 4, 1972 their "hunger march" on the capital was partially responsible for a coup which brought the populist military leader Oswaldo Lopez Arellano to power. He soon issued an emergency land reform measure, Decree Law 8. It was followed in January, 1975 by an agrarian reform law, Degree Law 170, which promised to distribute 600,000 hectares to 120,000 families in five years. While a substantial beginning was made in its implementation, the configuration of power moved significantly to its detriment when Lopez was removed from power in April 22, 1975 after news was revealed of a $1.25 million bribe paid by United Brands (United Fruit's new name) to the Honduran government in order to encourage (successfully) a substantial reduction in a proposed increase in the banana export tax. Reform-oriented officials were able to hold on to a share of power until the Spring of 1977, by which time most were removed, including Sandoval, who had been brought back in October, 1975 as the head of INA. Sandoval's position had been undercut from the beginning by the Chief-of-State's appointment as his subdirector an individual whose function was to restrain his boss (Ickis, 1983: 25). It should also be mentioned that resistance to the reform by large landowners stiffened during this period, including at times the use of force (Volk, 1981: 22-24).

An additional agrarian reform degree, Degree No. 78, was passed in September, 1981. Reflecting the efforts of the union of some 45,000 "very strong and politically conscious" coffee producers, this degree exempted coffee lands from the agrarian reform program and abolished the prior restrictions against the granting of titles through the reform process to holdings of under five hectares (US/AID, 1982: 5).

Several positive features of the Honduran agrarian reform should be noted. Perhaps most significant is the very fact that there has been, in a non-revolutionary situation, land redistribution and government intervention on the side of peasants in conflicts with larger landowners. Between 1962 and 1980, about 36,000 rural families benefited from the program, as indicated in Table 2.8. As Ruhl (1984: 55) points out, beneficiaries through the 1970s were equivalent to about 22 percent of the landless and land poor families in the mid-1970s.[15]

Despite its achievements, however, the Honduran agrarian reform has fallen far short of both its stated goals and the country's needs. As Table 2.8 shows, the peak of the reform process was reached in the 1973-1974 period by all three indicators---annual average number of families benefitted and land awarded, as well as the average size of the grants. The reform pace slackened after the promulgation of the 1975 law and by the end of 1980 the number of families benefited and the amount of land awarded were about one-sixth the stated goal of that law. In addition, the average size of the grants had declined. The reform pace quickened again when civilians returned to power under Roberto Suazo Cordova (1982-1985). However, the results were neither as impressive as those of the mid-1970s nor sufficient to stay even with the rapidly increasing population (Ruhl, 1985: 73-74). Consequently, Honduras now has more landless families than before the implementation of Degree No. 8 began in late 1972.

The usual form of the reform grants were to cooperative farms, apparently because of a belief on the part of government officials that larger farms were more conducive to agricultural modernization (Parsons, 1978: 10). The redistribution of land in Honduras was facilitated by the large quantity of public lands remaining in the country. Consequently, almost all of the land awarded was in fact public lands. Unfortunately, in many cases this land was in remote areas (US/AID, 1982: 4), insufficient, and of poor quality (Hatch, 1977: 13). While technical and credit assistance, including a $12 million loan initiated by AID in 1974, was made available to the cooperative farms as part of the reform process, the actual assistance received was often insufficient and difficult to obtain (Hatch, 1977; Tendler, 1976). For such reasons, the abandonment rate of reform farms has been high; by one estimate some 40 percent of the original settlers have abandoned such farms (US/AID, 1982: 4).

With the land redistribution program at a standstill, in the early 1980s the interest of both the Honduran and the U.S. governments switched to land titling.[16] In 1982, the two governments agreed to a Small Farmer Titling Project based on a $10 million loan and a $2.5 million grant from the United States. The intent of the project is to establish a viable system for granting fee simple property titles

Table 2.8
Land distribution under the reform process, 1962-1984

Period	Fam. Benft'd Total	Fam. Benft'd Annual Avg.	Land Awarded (hec) Total	Land Awarded (hec) Annual Avg.	Average Grant per Family[a]
1962-1966	281	56	1,357	271	4.8
1967-1972	5,348	891	34,604	5,767	6.5
1973-1974	11,739	5,870	79,552	39,776	6.8
1975-1977	12,405	4,135	80,150	26,717	6.5
1978-1981	9,174	2,294	38,937	9,734	4.2
1982-1984	13,241	4,414	58,770	19,590	4.4
TOTAL	52,188	2,269	293,370	12,755	5.6

[a] land grants often were to groups, not individual families. Therefore, this column does not give "average annual size of grant" but instead a measure of per family size of grants.

Sources: Calculated from James A. Morris, Honduras: Caudillo Politics and Military Rulers (Boulder, CO, 1984), and Mark J. Ruhl, "The Honduran Agrarian Reform Under Suazo Cordova", Inter-American Economic Affairs, vol. 39, no. 2 (Autumn, 1985), p. 70.

to 70,000 farmers. Special targets of the program are small and medium coffee growers; of the 48,000 coffee producers, about 95 percent have insecure titles to the land they farm (US/AID, 1982: 1-2). After the first two years of implementation, the program had provided 10,373 titles to small farmers, a rate pleasing to AID (US/AID, 1985: 116).

This program represents, as AID notes (p. 6), a change in philosophy on the part of the Honduran government away from supporting group farms "to an agrarian reform based on the principles of private property." While this new approach appears to offer nothing for the growing number of landless rural families, it does meaningfully address the real and serious problems of an important number of rural families. As previously indicated, almost no farmer in Honduras has a clear title to one's property. The 70,000 intended beneficiaries will gain an important measure of security through this program, especially the less powerful among them; it is estimated that about 57 percent of the intended beneficiaries work farm units of 10 hectares or less. The beneficiaries also gain better access to credit; a lack of title, necessary for establishing the right to use one's property for collateral, is a major impediment to obtaining commercial credit. Credit has become especially important in recent years as coffee rust has become a major threat to coffee growing. Control of the disease is possible but requires a substantial investment most onerous for small farmers (US/AID, 1982: 2-13).[17]

CONCLUSION

Agrarian reform has had a mixed history so far in Honduras. The first attempt at reform, the 1962 law, was largely both inspired by and aborted by external forces---elements of the U.S. government in both cases and the United Fruit Company in the latter. Even without the external forces, it is unlikely that the Honduran government had the power (or the will) to implement meaningful land redistribution. During the 1960s, however, substantial peasant mobilization altered somewhat the configuration of political power. Their dissatisfaction created new pressures on the government and new incentives for reformist policies by political leaders. As a consequence, policies of first the National Agrarian Institute in the late 1960s/early 1970s and then of the government under Lopez Arellano in 1973-1974 became more favorable to peasant interests. The zenith of populist reformism passed quicky, however, as government leadership changed and the opposition of large landowners solidified. It is also important to recall that little domestically-owned private land was redistributed; almost all of the land already had been public land, or was obtained from the fruit companies voluntarily or through expropriations in the less hospitable climate following the "Bananagate" disclosures of 1975 (Volk; 1981: 20-23).

Conditions were not as conducive to the cause of land redistribution in the first half of the 1980s.[18] Opposition by established interests continued to be substantial, outdoubtedly reinforced by the atmosphere generated by the war against Nicaragua. Politics based on mass mobilization and redistributive policies have always been difficult in Central America, but especially when their legitimacy could be questioned in the light of the vigilance claimed to be required by elites inface of the "communist threat" presented by Nicaragua. In such a climate, the creation of 10,000s of more secure small farmers was more feasible and probably politically astute. The switch in policy focus from land redistribution to titling certainly was in keeping with the policy orientation of the Reagan administration. Meanwhile, the number and percentage of landless and land poor rural families continued to increase while the per capita production of basic foods continued to decline.

A rapidly expanding population is a major cause of the plight of those rural families that have experienced increasing economic insecurity in the past few decades. This study also has documented, however, the significant role of the commercialization of agriculture, especially the expansion of export agriculture, as another major cause of declining access to land, food and income for many families. There is certainly much to commend an argument that in order to improve living conditions, Honduras needs to diversify and expand its exports, including agricultural exports (US/AID, 1984: 1). However, when such a policy is implemented under conditions of substantial inequality in the ownership of, and/or access to, resources, especially land, the experience in Central America indicates that the more probable result will be a decline in living conditions for many. This is especially true when prevailing patterns of land ownership create land scarcity. Some might point to the similar immiserization experienced by rural peoples during the spread of commercial agriculture elsewhere, such as in Great Britain, and argue that such trends appear to be intrinsic to the developmental process, but will lead to improved living conditions for subsequent generations, if not all of those in the present. In addition to its normative assumptions, this position must be contested on the basis of the evidence that has developed thus far for Central America. In the light of this evidence we must conclude that the rural development policies of the Honduran and United States governments, especially in the 1980s, are based more on ideology, wishful thinking, and self-interest, than they are on critical analysis.

NOTES

*An earlier version of this study was presented at the 1985 Annual Meeting of The American Political Science Association.

1. They also estimate that even with a high rate of economic growth through 1990, 41 percent would still consume at least 10 percent less than recommended (Reutlinger and Alderman, 1980: Appendix Table 2A).

2. More generally, see de Janvry (1981), Feder (1976), Grindle (1986), Hewitt de Alcantara (1976), and Pearse (1980).

3. In 1954 banana workers, in a major surprise, went on a prolonged strike which had important ramifications for national politics (Anderson, 1981; MacCameron, 1983).

4. Symbolic of the relationship of these companies to the rest of the country is the story of the railroads they did not develop. Much of their land was obtained free on the promise of constructing badly needed railroads. Hondurans expected railroads which would serve the interests of their country; instead of linking the coast with major cities in the interior, banana companies stopped with lowland railroads which served their needs instead (Kepner and Soothill, 1967: 113-115, 147-152).

5. A more accurate estimate of the supply of basic food, but for a far shorter period of time, is possible using data supplied to the U.S. Department of Agriculture by its Honduran attache. When imports, exports, livestock feed use, and stocks on hand at the beginning of the year are taken into account for the mid-1970s, the effect on per capita food supply is negligible.

6. This conclusion is reinforced by other studies. Using essentially the same starting point (approximately 1960), the Central American Bank estimated per capita meat consumption to have declined 24 percent by the mid-1960s (Fonck, 1972: 57); Morgan (1973: 6) estimates the decline to 1972 to be 13 percent; Slutzki (1979: 114) estimates the decline to 1975 to be 35 percent. Slutzki's data is presented on an annual basis, thereby allowing a comparison with the other studies; his data for each period substantially agree with the others.

7. There is evidence that cattle are shipped clandestinely to Guatemala to take advantage of its higher beef prices. Similarly, the border with Nicaragua, at least prior to 1980, was also porous (USDA/FAS, 1979; 1980).

8. In 1984, the U.S. government reduced the Honduran beef quota for that year, causing the closing of seven meat exporter companies and the loss of jobs of some 3000 workers (Honduran Information Center, 1984: 8).

9. Ruhl (1984) gives an excellent comparison of land ownership and tenure patterns between Honduras and El Salvador; for Guatemala see Brockett (1984).

10. At the other end of the scale, land held by the largest farms declined by six percent, most of which shifted into the next largest category.

11. In fact, too much so; land scarcity in Central America leads to over-use of the land and, therefore, declining fertility and environmental degradation, such as erosion and deforestation.

12. The project proposals usually call for pasture land to be reduced to about 20 percent of the total.

13. A study of the government's efforts of the first years concluded that it had added perhaps five percent to production (Checchi, 1959: 62).

14. Land scarcity and competing claims had much to do with the major event of this period, the so-called 'soccer war' with El Salvador in 1969 (Anderson, 1981).

15. Ruhl (1984:55) also notes that the reform "was very important symbolically because the program demonstrated the

continued flexibility and reform potential of the Honduran government and fostered an 'incrementalist' policy orientation among the peasant organizations."

16. This is, of course, in step with the evolution of the agrarian reform program in El Salvador, with the change from redistribution to "land to the tiller" (Simon, 1982).

17. In 1981, the United States initiated a five-year $9.55 million program in Honduras to help coffee growers, including to combat coffee rust.

18. For a more complete discussion see Ruhl (1985).

REFERENCES

Anderson, Thomas P. (1981). The War of the Dispossessed: Honduras and El Salvador, 1969. Lincoln: University of Nebraska Press.

Bakken, Henry H., et al. (1965). Marketing and Storage Facilities for Selected Crops: Honduras. Kansas City, Mo.: Weitz-Hettelsater Engineers.

Blutstein, Howard I., et al. (1971). Area Handbook for Honduras. Washington, D.C.: U.S. Printing Office.

Brockett, Charles D. (1984). "Malnutrition, Public Policy and Agrarian Change in Guatemala," Journal of Inter-American Studies and World Affairs, 26, 4 (November): 477-497.

_____. (1985). "The Right to Food and International Obligations: The Impact of U.S. Policy in Central America." In George W. Shephard and Ved P. Nanda. eds. Human Rights and Third World Development. Westport, CN: Greenwood Press, 1985.

Burrows, Charles R. (1962). Letter to Edward M. Rowell, Department of State, November 18, Presidential Office Files, Countries, Box 18, John F. Kennedy Library.

_____. (1969). Recorded interview by Dennis J. O'Brien, Sept. 4, John F. Kennedy Library Oral History Program.

Checchi, Vincent (1959). Honduras: A Problem in Economic Development. New York: Twentieth Century Fund.

de Janvry, Alain (1981). The Agrarian Question and Reformism in Latin America. Baltimore: The Johns Hopkins University Press.

Durham, William H. (1979). Scarcity and Survival in Central America: Ecological Origins of the Soccer War. Stanford: Stanford University Press.

Feder, Ernst (1976). "How Agribusiness Operates in Underdeveloped Agricultures: Harvard Business School Myths and Reality," Development and Change, 7 (October): 413-443.

Fonck, Carlos O. (1972). Modernity and Public Policies in the Context of the Peasant Sector: Honduras as a Case Study. Ithica: Unpublished doctoral dissertation, Cornell University.

Grindle, Merilee S. (1986). State and Countryside: Development Politics in Latin America. Baltimore: The Johns Hopkins University Press.

Hatch, John K. and Aquiles Lanao Flores (1977). An Evaluation of the AIFLD/HISTADAUT Project Proposal To Assist Peasant Federations. Prepared for U.S. Agency for International Development. Washington, D.C., August 19.

Hewitt de Alcantara, Cynthia (1976). Modernizing Mexican Agriculture. Geneva: United Nations, UNRISD.

Honduras Information Center (1984). "Chronology," Honduras Update, 2, 5 (February): 8.

Honduras, Republic of (1954). Primer Censo Agropecuario Tegucigalpa: Ministerio de Gobernacion.

Ickis, John C. (1983). "Structural Responses to New Rural Development Strategies." In David C. Korten and Felipe B. Alfonso. eds. Bureaucracy and the Poor. West Hartford, Conn.: Kumarian Press.

IDB (1978; 1984). Economic and Social Progress in Latin America. Washington, D.C.: InterAmerican Development Bank.

Keene, Beverly (1980). "Export-Cropping in Central America" Washington, D.C.: Bread for the World, Background Paper, No. 43.

Kepner, Charles David Jr. and Jay Henry Soothill (1967, reissue of 1935). The Banana Empire: A Case Study of Economic Imperialism. New York: Russell and Russell.

LaFeber, Walter (1984). Inevitable Revolutions: The United States in Central America. New York: W.W. Norton and Company.

MacCameron, Robert (1983). Bananas, Labor, and Politics in Honduras, 1954-1963. Syracuse, NY: Maxwell School of Citizenship and Public Affairs.

Molina Chocano, Guillermo and Ricardo Reina (1983). La Evolucion de la Pobreza Rural en Honduras. Santiago, Chile: International Labor Organization, PREALC/223, March.

Morgan, Q. Martin (1973). The Beef Cattle Industries of Central America and Panama. Washington, D.C.: U.S. Department of Agriculture, Foreign Agricultural Service.

Morris, James A. (1984). Honduras: Caudillo Politics and Military Rulers. Boulder, Co.: Westview Press.

Nesman, Edgar G. (1981). Peasant Mobilization and Rural Development. Cambridge, MA: Schenkman Pub. Co.

Parson, Kenneth H. (1976). Agrarian Reform in Southern Honduras. Madison: University of Wisconsin Land Tenure Center Research Paper, No. 67, March.

_____. (1978). Key Policy Issues for the Reconstruction and Development of Honduran Agriculture Through Agrarian Reform. Madison: University of Wisconsin Land Tenure Center, No. 114, January).

Pearse, Andrew (1980). Seeds of Plenty, Seeds of Want. Oxford: Claredon.

Pearson, Neale J., (1980). "Peasant Pressure Groups and Agrarian Reform in Honduras, 1962-1977." In William P. Avery, et al. ed. Rural Change and Public Policy. New York: Pergamon Press.

Perez Brignoli, Hector (1983). "Growth and Crisis in the Central American Economies, 1950-1980," Journal of Latin American Studies, 15 (November): 365-398.

Pfeil, Ulrike (1977). Peasant Mobilization and Land Reform: A Theoretical Model and a Case Study (Honduras). Gainesville: Unpublished master's thesis, University of Florida.

Reutlinger, Shlomo and Harold Alderman (1980). "The Prevalence of Calorie-Deficient Diets in Developing Countries," World Development, 8.

Ruhl, J. Mark (1984). "Agrarian Structure and Political Stability in Honduras," Journal of Interamerican Studies and World Affairs, 26, 1 (February): 33-68.

_____. (1985). "The Honduran Agrarian Reform Under Suazo Cordova," Inter-American Economic Affairs, 39, 2 (Autumn): 63-81.

Shane, Douglas R. (1980). Hoofprints on the Forest. Washington, D.C. Study prepared for U.S. State Department.

Shaw, Royce Z. (1979). Central America: Regional Integration and National Political Development. Boulder, Co.: Westview Press.

Simon, Laurence R. and James C. Stephens, Jr. (1982). El Salvador Land Reform, 1980-1981: Impact Audit. Boston: Oxfam America.

Slutzki, Daniel (1979). "Agroindustria de la carne en Honduras," Estudios Sociales Centroamericanos, 7, 22: 101-205.

Staatz, John M. and Carl K. Eicher, (1984). "Agricultural Development Ideas in Historical Perspective." In Carl K. Eicher and John M. Staatz. eds. Agricultural Development in the Third World. Baltimore: Johns Hopkins University Press.

Tendler, Judith (1976). *Intercountry Evaluation of Small Farmer Organizations: Final Report (on) Ecuador and Honduras.* Washington, D.C.: U.S. Agency for International Development Program Evaluation Studies.

US/AID (1982). *Honduras, Project Paper: Land Titling.* Washington, D.C.: Agency for International Development.

———. (1983). *Honduras, Project Paper: Small Farmer Livestock Improvement.* Washington, D.C.: Agency for International Development.

———. (1984). *Honduras, Project Paper: Export Promotion and Services.* Washington, D.C.: Agency for International Development.

———. (1985). *Congressional Presentation, Fiscal Year 1986, Annex 3. Latin America and the Caribbean, Volume 1.* Washington, D.C.: Agency for International Development.

US/Congress (1962). *Congressional Record.* Washington, D.C., U.S. Printing Office, October 2.

USDA/FAS (1979, 1980, 1981). *Attache Report: Honduras - Livestock Meat.* Washington, D.C.: U.S. Department of Agriculture, Foreign Agricultural Service, Nov. 6, 1979, June 5, 1980.

US/GAO (1970). *Administration and Effectiveness of U.S. Economic and Military Aid to Honduras.* Washington, D.C.: U.S. Central Accounting Office, Dec. 3.

US/House, Committee on Foreign Affairs, Subcommittee on Interamerican Affairs, (1979). *Foreign Assistance Legislation for FY 1980-1981, pt. 5,* 96th Cong. 1st Sess., February 20.

Volk, Steven (1981). "Honduras: On the Border of War," *NACLA Report on the Americas,* XV, 6 (November-December): 2-37.

Weeks, John (1985). *The Economies of Central America.* New York: Holmes and Meier.

World Bank, et al. (1978). *Honduras: Agricultural/Rural Sector Survey Volume IV.* World Bank Working Paper.

3

Brazilian Agriculture and the Debt Crisis

Bruce Drury

Developing nations of today's world see their future prosperity in the establishment of modern industry. An expanding and diversified industrial base will provide the jobs, materials and capital needed for the transition to a truly modern service-oriented society. Brazil, a nation in which agriculture has traditionally been dominant, has made significant progress in its quest for industrialization. Consumer goods, heavy machinery, automobiles, ships, electronics and weapons systems are made in Brazil for domestic use and export.

Agriculture is still very important to Brazil. Although agriculture's share of the total GDP is only 10 percent, 36 percent of the economically active population work in the agricultural sector. Furthermore, agriculture in 1984 generated 41 percent of the country's exports while food imports accounted for only 9 percent of the total imports. Nearly half (48 percent) of Brazil's imports were financed by agriculture (FAO, 1985: 164).

Although the emphasis has been on industrialization, Brazilian politicians recognize the importance of the agricultural sector. They have discussed the necessity of modernizing agriculture in a manner that would increase food production for domestic consumption and export, alleviate economic inequities in rural Brazil and increase land access and rural employment opportunities. Several governmental initiatives---the Land Statute of 1964, the Program for National Integration, PROTERRA and others---were, although conservative in nature, significant efforts toward agrarian reform. Unfortunately, Brazil's huge external debts created the short-term necessity to generate a trade surplus through increased agricultural exports and thus effectively canceled the reform programs. International economic pressure reinforced the already significant political power of the latifundists and large scale commercial farmers, and the result has been the surrender of long-term, idealistic objectives in favor of short-term political and economic reality. Export agriculture has defeated food and reform.

THE ROLE OF AGRICULTURE

Although Brazil is rapidly industrializing, agriculture is still the largest economic sector of the society. It is the provider of food for a rapidly growing urban population, the major generator of foreign exchange and a major market for domestically produced machinery and consumer goods. The importance of the agrarian sector was expressed in the government's three-year Plan for Economic and Social Development, 1963-1965, by stating that "...the country's deficient agrarian structure...stands as the most serious obstacle to a rational improvement...of our domestic economy" (quoted in Syvrud, 1974: 216). More recently Brazil's Third National Development Plan, 1980-1985, assigns to agriculture the tasks of reducing inflation by increasing the supply of food, earning foreign exchange through agricultural exports, reducing oil imports through the production of alcohol and creating rural jobs in order to reduce poverty and rural-to-urban migration (World Bank, 1982: 3).

The agricultural sector has been quite successful in producing foreign exchange (but not without a significant cost), and it has produced sufficient alcohol to allow a significant reduction in oil imports. Furthermore, the expansion of export agriculture has helped the industrial sector. Jose Montello of the Brazilian Institute of Geography and Statistics reported that the five percent increase in industrial output in the first half of 1984 was due almost entirely to the demand from Brazilian agriculture for machinery, fertilizer and chemicals (O Estado de Sao Paulo, 16 August 1984: 36).

These successes have incurred a price, however, because Brazilian agriculture is not yet capable of performing simultaneously all the tasks outlined in the Third National Development Plan. It can not deliver foreign exchange surpluses and, at the same time, produce and distribute sufficient food for the populace. Even before export agriculture became the major emphasis, malnutrition was a critical problem in Brazil, especially in the cities. In 1975, it was estimated that only 32.8 percent of the populace had an adequate diet. 18.6 percent had deficits of up to 200 calories per day, 31.3 percent had deficits of 200 to 400 calories and 17.3 percent had deficits of more than 400 calories (Knight and Moran, 1981a: 92). Brazil's infant mortality rate in 1978 was almost twice that of nine other developing nations with equivalent GNP's (Knight and Moran, 1981a: 21). Although the government has used price controls and has maintained a rather expensive program of subsidies for certain foods, it has not been successful in thwarting malnutrition and related health problems.

The debt crisis and the emphasis on export agriculture have seriously exacerbated this "social debt." Food prices rose rapidly as land and other resources were directed away from food production. Thus, more and more Brazilians were unable to afford an adequate diet, and health officials estimate that the caloric intake of Brazilians declined by 18 percent from 1967 to 1981 (Latin American Regional Report-Brazil, 1981: 7).

The shift to large scale export agriculture not only altered food availability but also affected employment availability. Some peasants lost rental properties as commercial operators expanded their farms to make optimal use of large machinery. Many farm workers lost jobs as machines replaced humans in the fields. Without secure employment opportunities, many rural Brazilians are migratting to the cities in hope of finding work.

Just as the agricultural sector has been seen as a cause of many social and political problems, reform of the rural social and economic structure has been tagged as a cure for those ills. The inequity of the land tenure system creates significant economic and political power for the conservative latifundists, deprives Brazilian peasants of the opportunity to make an adequate living and foments unrest in the countryside and the cities. Government programs which were aimed at equalizing land and credit availability have twin goals of increased productivity and the alleviation of many social problems created by the existing inequities. For example, the 1975 National Alcohol Program (PNA) had the primary goal of reducing Brazil's reliance on imported petroleum, but it also had the social objectives of reducing regional and individual income disparities and of expanding employment opportunities in agriculture (Saint, 1982: 224). Unfortunately, the social goals of the PNA and other programs have been frustrated, in part, by the debt crisis.[1]

With abundant land and labor and a generally favorable climate, Brazil has the potential for feeding her own people quite well and of producing significant surpluses for export. However, various factors relating to debt crisis have created rather extreme distortions in the agricultural sector and have in turn resulted in severe political difficulties for the government.

POST-COUP AGRICULTURAL EXPANSION, 1964-1973

In spite of the fact that Brazil was a predominantly agrarian nation, agriculture received relatively little governmental attention (except for the Northeast during that region's periodic droughts) until the military coup in 1964. In the post-WW II period Brazilian farm production expanded steadily in spite of policies which gave preferences to industry and actually discriminated against the agricultural sector. An overvalued exchange rate discouraged exports and price controlson basic commodities inhibited the expansion of food production for internal consumption. Low factor costs, such as abundant and cheap labor and inexpensive land on the frontier and an expanding highway network were the impetus for increased production (WorldBank, 1982: 2).

The military government eased price controls in 1964, liberalized trade policy in 1967 and established a moving and generally realistic exchange rate. These factors, plus a general expansion of world commerce in the late 1960s and early 1970s, resulted in a significant increase of agricultural production.

As an authoritarian regime without electoral legitimacy, the military government attempted to legitimize itself through economic performance. Industrial expansion was the major area of emphasis, but agricultural policy was rationalized and the government also attempted to stimulate sustained agricultural growth through governmental initiatives that were designed to have desirable social consequences. The 1964 Land Statute (Estatuto da Terra) established a system of progressive taxation on large and inefficient farms and set up a program for the distribution of government land to landless peasants (Kutcher and Scandizzo, 1981: 196-197). Since production per hectare varies indirectly with the size of the farm (small farms receive proportionally greater inputs of capital and labor), the Land Statute should have increased production while reducing economic inequities (Berry and Cline, 1979: 44-53; Kutcher and Scandizzo, 1981: 83-87) if the law had been enforced. As it turned out,the reform-by-taxation was defeated by low original evaluations of the unused land, by inflation which took the sting out of the tax, and by the fact that the tax was determined by state authorities who were often sympathetic to the latifundist landowners.

Similar efforts to stimulate rural development, especially in the Northeast and the Amazon, were made through the Program for National Integration (PIN) of 1970 and PROTERRA of 1971. PIN envisioned a vast colonization scheme in the Amazon. Sixty thousand families were to be provided with one hundred hectare farms at minimal cost and were to receive extensive financial and technical assistance (Mahar, 1979: 22-23). PIN had set up only six thousand families, most of whom left their farms when the promised government support did not materialize. The program failed because the planners did not understand the necessity for special crops and cultivation techniques in the Amazon and because Northeast landowners and industrialists opposed any project that would drain away capital and cheap labor. PROTERRA was created to buy and distribute land in the Northeast and again the government was to provide considerable capital and agricultural extension support (Kutcher and Scandizzo, 1981: 17-18). PROTERRA had distributed only seventeen thousand hectares of land when the debt crisis altered the policy orientation of the government. Poor planning, opposition from the landowners, agency infighting, and finally, the debt crisis defeated these efforts to increase productivity and to reduce individual and regional inequities (Katzman, 1977: 144-146). Ultimately, the government used the reform programs to justify investments in expensive infrastructure that primarily benefit large scale commercial enterprises oriented toward exports (Schmink, 1982: 343-344; Bunker, 1984: 95-96).

Brazil's "economic miracle" of the late 1960s and early 1970s was the result of the government's use of repressive measures to hold down labor costs and to reduce inflation. This then allowed Brazilian industry to expand very rapidly by utilizing unused capacity in an atmosphere of liberal trade policy and an expanding world economy. Agricultural performance during this period was not "miraculous" but it did respond positively to increased credits for farmers, limited price regulation, and the free trade atmosphere. Unfortunately, these policies were largely discontinued after 1973 and, more importantly, the reform and colonization programs which could have been the basis for a more prosperous and egalitarian rural sector were either abandoned or reoriented in favor of export agriculture.

THE DEBT CRISIS

President Jose Sarney, the Brazilian government, and Brazil have the unpleasant task of paying the interest on a foreign debt of over $100 billion and ultimately they must find a way of paying off the principal.

Brazil's debt crisis stems from the post-WWII decision to emulate the U.S. model of petroleum-based, energy intensive development. This decision then led directly to three corollary decisions. First, Brazil opted to create PETROBRAS, a vertically-integrated, state monopoly oil company, to exploit, refine, and distribute the oil that every Brazilian believed existed under the country's vast surface. Second, the government chose to expand the transportation system by means of a highway network and, in turn, to allow the rail and maritime systems to atrophy. Third, the cornerstone of the industrial sector would be the manufacture of the vehicles which would burn the PETROBRAS oil on the new highway system (Smith, 1984: 3). Alas, the expected oil eluded discovery and the result has been huge debts, unequal distribution of wealth, malnutrition, increased infant mortality, and a variety of other negative economic and social indicators (Erickson, 1981: 164-165).

The folly of the development model did not become apparent until after 1973 because, although Brazil did not find the expected oil, the international price of oil was low and the nation accepted oil importation as a temporary and cheap inconvenience. Even after the first oil shock in 1973, policy makers were not overly alarmed because there was an abundant supply of loan money for imported oil and for a number of megaprojects---petro-chemical plants, aircraft factories, nuclear power plants---whose high tech exports or replacement of imports would ultimately allow Brazil to pay off the still moderate debt (Knight and Moran, 1981b: 22-23).

The technocrats in the government responded to the first oil shock with supply-oriented policies that sought to provide adequate energy to the energy intensive industrial sector. This not only added to growing debt but the emphasis on capital projects diverted attention and money from smaller, less glamorous, labor intensive enterprises which could have taken advantage of Brazil's labor surplus (Knight and Moran, 1981b: 23). When the second shock hit in 1979 as a result of the revolution in Iran, Brazil was forced to restrict demand through price increases, which affected the public far more than the protected industries (Erickson, 1981: 168-169). Inflation, higher food prices, unemployment, service cuts, tax increases, and reductions in real wages were the result of the government's attempt to maintain a supply of capital and energy for the high-tech megaprojects.

The second oil shock was accompanied by an increase in interest rates which made a "debt problem" a true "debt crisis." A Brazilian scholar wrote:

> At first, between 1970 and 1976, Brazil borrowed money to increase its rate of accumulation and consumption. Later, between 1978 and 1980, Brazil borrowed money to maintain its level of consumption. Since 1981, we borrow money not to maintain consumption, but almost exclusively to pay interest (Pereira, 1984: 172).

Celso Furtado contends that Brazil and the international banking community were equally to blame for the crisis. Brazil was eager to borrow and the bankers were eager to lend until things got out of hand in the 1979-1982 period. Deteriorating terms of trade and higher interest rates cost Brazil an estimated $47 billion, more than one half of its 1982 debt. The bill for interest alone in this period was $31 billion (Furtado, 1984: 3).

By 1982, Brazil had two alternatives. She could declare a complete moratorium (i.e., default) and risk becoming a pariah nation in financial circles. This would have put virtually all of the development projects in jeopardy. The other alternative was to work through the International Monetary Fund in an effort to restructure the debt. Brazil's avoidance of the IMF dates back to the 1950s when President Kubitschek denounced the IMF as the agent of international banks which was incapable of understanding the problems of developing nations (Pereira, 1984: 206-207). This time, however, Brazil had little choice. Negotiations with the IMF in 1982 resulted in an agreement giving Brazil a temporary moratorium on the principal and $7 billion in new loans in exchange for a Brazilian promise to reduce inflation, government spending, and imports and to produce a $6 billion trade surplus in 1983.

The debt crisis has been managed but certainly not overcome. Brazil did surpass the balance of trade requirement in 1983 by ending the year with a surplus of $6.4 billion. The 1984 surplus was nearly $13 billion and, in spite of the world-wide recession, Brazil ended 1985 with a surplus of $11 billion. Brazil's performance has allowed payment of the interest (which will be nearly $15 billion in 1987) and has persuaded the international banking community to reschedule and stretch out payments on the principal. The austerity measures of the IMF and the world-wide recession resulted in a severe recession in Brazil. The economy can not be squeezed much more without massive unrest that could topple the new democratic government. The major squeeze has been on agriculture and food. Regressive agricultural policy has resulted in massive unemployment in rural areas and too little food in the cities.

EXPORTS VERSUS FOOD AND REFORM

Brazil certainly has the potential to take advantage of export opportunities, as an American economist stated in 1975:

> The most promising future for the agricultural sector is an expansion of exports. In a world where food shortages seem to be evermore serious, Brazil occupies the unique position of having almost infinite possibilities for expanding agricultural output through adding acreage and through increased productivity (Robock, 1975: 106).

Given the error in the choice of the development model, it was indeed fortunate that Brazil had the ability to take advantage of the agricultural export market.

After the first oil shock in 1973, the government established a clear bias for industry over agriculture because the country could not afford to lose its investment in industrialization. Within agriculture there was a clear bias in favor of exports and import substitution as credit subsidies went predominantly to large commercial farming enterprises (World Bank, 1982: 2-3). A shift from domestic food crops to export crops was a relatively easy way to quickly increase the export earnings needed to buy oil. The value of agricultural exports in constant U.S. dollars increased from $4.9 billion in 1973 to an estimated $14.7 billion in 1984 (World Bank, 1982: 13; <u>O Estadode Sao Paulo</u>, 7 Jan 1984: 23) with agricultural exports accounting for more than 60 percent of Brazil's total export value. The increases for six major export commodities are shown in Table 3.1. The rather stable value of sugar exports is remarkable given the fact that a large proportion of the sugar production was, by 1984, being used in import substitution for the production of fuel alcohol.

Table 3.1
Value of Major Agricultural Exports, 1973, 1981-1984[a]
(in millions of U.S. dollars)

Commodity	1973	1981	1982	1983	1984
Beef	215	414	443	515	521
Chicken	negligible	354	286	242	260
Orange Juice	64	659	573	609	1,415
Sugar	527	975	600	515	586
Coffee	1,244	1,517	1,858	2,325	2,856
Soybeans	527	3,191	2,122	2,562	2,566
TOTAL	2,577	7,110	5,882	6,768	8,204

[a]The values expressed include basic commodities and processed products.

Source: IBGE, various years. 1984 data from Banco Central do Brazil (1985: 377).

Export earnings from agriculture were obviously beneficial in managing the debt problem, but the tilt toward export agriculture had a deleterious effect on the distribution of land and income in the rural sector. In Amazonia, the small farm colonization aspect of the PIN was shelved in 1975 in favor of support for highly capitalized ranching and mining enterprises, which would generate more foreign exchange. Many small farmers, either purchasers or squatters, were then evicted from their land by legal or violent means (Bunker, 1984: 1050). Similar situations occurred throughout the country as the commercial enterprises consolidated large holdings at the expense of small-farm, food producers.

The government's support for export agriculture was largely limited to production credits, but the large amounts of money provided to the commercial farming enterprises had a major impact. The National System of Rural Credit, established by Getulio Vargas in 1937 as a means of assisting small farmers, directed most of its credits to the large producers. In 1979, 10 percent of the recipients received 72.8 percent of the loans (<u>Latin American Regional Reports: Brazil</u>, 7 March 1981: 5). Since the interest rate was generally only one-fifth or less than the inflation rate, the commercial farmers often put the money in high paying commercial savings accounts, or they use it in land speculation. Food crop producers were deprived of capital and could not compete for land with land values inflated by the speculation. In spite of the surpluses in export agriculture, Brazil had to import significant amounts of food (Smith, 1984: 14-15).

The government's preferential treatment of large-scale export agriculture made it difficult for the food producers to secure modern inputs of fertilizers and pesticides because they lacked the credit needed to buy these commodities, which had become very expensive because of higher petroleum costs (World Bank, 1982: 57-59). Furthermore, the debt problem was made more severe because the commercial operators require more petroleum fuel and petroleum-based fertilizers and chemicals than do the small holders.

As a result of the government's benevolence for export agriculture, food production in Brazil has remained relatively flat while the production of export commodities, as shown in Tables 3.2 and 3.3, has increased dramatically. Table 3.2 gives the aggregate production of eight major farm crops for 1970 through 1983. Table 3.3 measures the change in production of these crops as a percentage of the average production for 1970, 1971 and 1972. The three-year average is an attempt to overcome the variability created by weather and other short term factors. Although sugar, soybeans (as edible oil) and coffee are consumed in Brazil, they are primarily export crops, or as in the case of sugar, for import substitution. Sugar cane and soybeans clearly exceed food crops of rice, manioc, black beans, wheat and corn. Since about 80 percent of the corn is used for animal feed (World Bank, 1981: 12), its inclusion as a domestic food crop is questionable. The increase in corn production closely parallels the increase in exports of beef and frozen chickens.

Figure 3.1 shows production changes in food and export crops in relation to population. The export crops---sugar, soybeans and coffee---make impressive increases from 1976 to 1983 and stood, in 1983, at 272 percent of the 1970-1972 average. On the other hand the food crops of rice, manioc, wheat, beans and corn did not keep up with population increase. In 1983 the five food crops were only 94 percent of 1970-72 production while the population had increased by

Table 3.2
Production figures for major agricultural crops, 1970-1983
(in thousands of metric tons)

Year	sugar	soybeans	coffee	rice	manioc	black beans	wheat	corn
1970	79,753	1,508	1,510	7,553	29,465	2,211	1,844	14,216
1971	80,838	2,077	3,103	6,593	30,229	2,688	2,011	14,130
1972	85,106	3,223	1,410	7,824	29,830	2,676	684	14,500
1973	91,994	5,012	1,746	7,160	26,527	2,332	2,031	14,186
1974	95,624	7,876	3,231	6,764	24,798	2,238	2,858	16,273
1975	91,525	9,893	2,545	7,782	26,118	2,282	1,788	16,334
1976	103,173	11,223	752	9,757	25,443	1,840	3,212	17,751
1977	120,082	12,513	1,951	8,894	25,929	2,290	2,066	19,256
1978	129,145	9,541	2,535	7,296	25,495	2,194	2,691	13,569
1979	138,899	10,240	2,666	7,595	24,926	2,186	2,927	16,306
1980	146,290	15,153	2,133	9,748	24,045	1,969	2,641	20,374
1981	155,924	15,007	4,064	8,228	24,516	2,341	2,210	21,117
1982	184,219	12,835	2,007	9,718	24,039	2,907	1,820	21,865
1983	216,703	14,582	3,331	7,750	21,746	1,587	2,265	18,744

Source: IBGE, (various years).

Table 3.3
Changes in agricultural production of major crops, 1973-1983
(index 1970-1972=100)

Year	sugar cane	soybeans	coffee	rice	manioc	black beans	wheat	corn	population
1970-1972	100	100	100	100	100	100	100	100	100
1973	112	220	87	98	89	92	134	99	105
1974	117	347	161	92	83	87	189	114	108
1975	112	430	127	106	88	90	118	114	111
1976	126	495	37	133	85	73	212	124	115
1977	145	551	97	123	87	91	137	135	118
1978	158	420	126	100	85	87	178	95	121
1979	170	451	133	104	83	87	193	114	125
1980	179	668	106	133	81	78	175	143	128
1981	190	661	202	112	82	93	146	148	129
1982	225	566	100	133	81	115	120	153	132
1983	265	643	166	106	73	63	150	131	135

Source: Derived from Table 3.2.

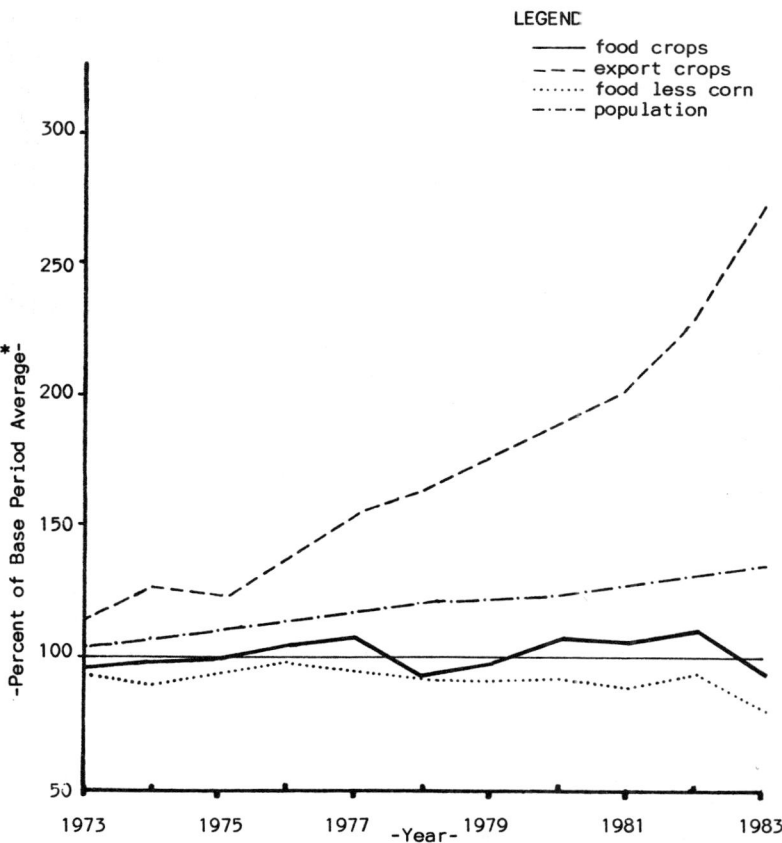

*base data are averages of 1970-1972

Figure 3.1 Changes in Brazilian Agricultural Production and Population: 1973-1983

Source: Computed from Table 3.2

33 percent. If corn is eliminated as a food crop because of its use as rations for animals intended for export, major food crop production declines to only 81 percent of the 1970-72 average.

After the extremely poor performance of food crops in 1978 and 1979 (partly due to weather conditions), the government increased its credits for agriculture. Wheat and rice farmers received twice as much money in the form of subsidized loans as their production proportions should have indicated, but two other important food crops, black beans and manioc, were almost ignored (World Bank, 1982: 20-23). Even though more money went to wheat and rice producers, the bias was still in favor of export agriculture and relatively little money went to small holders. Credit for the wealthy few lowered crop prices for all, increased the competition for land and allowed profit-taking from the commercial loan market (World Bank, 1982: 24-25).

Recognizing the poor performance of food crops, Agriculture Minister Nestor Jost set 1985 planting targets for food crops at two million hectares over 1984. But he was unwilling to promise that adequate credits for food crops would be available for 1985 even though he admitted that the government's bias for export crops was a major problem in food production (O Estado de Sao Paulo, 2 June 1984: 23).

Government efforts to control inflation and to satisfy the people's taste for bread and wheat products targeted wheat for special subsidies and price controls. The efforts backfired, however, as the price controls increased the demand for wheat at the expense of manioc and other grains, while weather conditions and inappropriate farming practices kept production down (World Bank, 1982: 42-44). Over the past decade Brazil has normally imported over one half of its wheat consumption. Brazil did increase production by one hundred thousand tons in 1983, not as a result of government encouragement but as a result of higher yields due to favorable weather. Land once used for wheat in Rio Grande do Sul and Parana continues to be diverted to higher priced export crops (O Estado de Sao Paulo, 5 Jan 1984: 26). The cost of wheat imports has declined from about one billion dollars in 1978-79 to a little more than $800 million in 1984 only because of government restrictions on imports.

Although export crop production has expanded, production increases have occurred because of extensive measures, not intensive practices. Agricultural exports in 1984 registered an increase of more than one billion dollars over 1983 but the increase was due to additional planting and especially to higher prices for coffee, cocoa, orange juice and soybeans (Journal do Brasil, 11 June 1984: 16). As shown in Table 3.4, land planted in food crops increased by only 6 percent from 1973 to 1983 while land devoted to export crops increased by 81 percent.

Productivity per hectare has not improved and yields in 1983 for two of the crops was actually down significantly in comparison to 1973, as demonstrated in Table 3.5. The rather unimpressive performance in yields generally reflects the extension of farming into new, marginal land and the low application of modern inputs such as fertilizer, pesticides and improved varieties (Baer, 1983: 319). In general, the producers of export crops have received more government credits and thus could purchase better seed, fertilizers and pesticides, and they also were able to secure the better land. Food producers were pushed to the marginal land and could not buy modern inputs.

Animal production in Brazil has not been impressive in spite of the vast territory devoted to ranching. The one area of significance is poultry and here again production for export is favored. Brazil has found a favorable market for frozen chickens, as was shown in Table 3.1, and this in turn explains a significant part of the increase in maize production. Beef exports have increased, as indicated in Table 3.1, but at the expense of domestic consumption. Inflation and reduced purchasing power has forced Brazilians to switch to cheaper pork and poultry, and thus more beef is available for export (Latin American Regional Report: Brazil, 19 Oct 1984: 7). Again the need to generate foreign exchange comes ahead of increased domestic food supplies.

IMPORT SUBSTITUTION VERSUS FOOD

Import substitution is a common practice used by developing nations to promote domestic industry and to reduce the drain on foreign exchange. Given Brazil's relative lack of domestic petroleum, the increasing price of imported oil, and the already sizable debt, the 1975 National Alcohol Program (PNA) made good sense. The technology had already been developed in Brazil and, besides the balance of payment benefits, alcohol production from cane sugar or manioc could be quite favorable to small farmers, especially in the impoverished Northeast. As initially presented, the PNA would not detract from food production and would have a variety of positive economic and social aspects.

Table 3.4
Area planted in major crops in thousands of hectares, with 1973 to 1983 (change in percentage)

Crop	1973	1976	1979	1981	1982	1983	% Change 1973-1983
rice	4,795	6,656	5,452	6,102	6,016	5,110	107
manioc	2,104	2,094	2,111	2,067	2,126	2,021	96
black beans	3,815	4,059	4,212	5,027	5,930	4,069	107
wheat	1,839	3,540	3,831	1,920	2,625	1,885	103
corn	9,908	11,118	11,319	11,520	12,601	10,742	108
FOOD TOTAL	22,461	27,467	26,925	26,636	29,298	23,827	106
sugar cane	1,959	2,093	2,537	2,826	3,073	3,447	176
soybeans	3,615	6,417	8,256	8,501	8,202	8,136	226
EXPORT TOTAL	7,654	9,631	13,199	13,945	13,132	13,862	181
TOTAL	30,115	37,098	40,104	40,581	42,430	37,689	125

Source: IBGE (various years).

Table 3.5
Yield of major crops in Kg. per hectare with 1973 to 1983 change in percentage

Crop	1973	1976	1979	1981	1982	1983	% Change 1973-1983
rice	1,495	1,466	1,393	1,348	1,615	1,517	101
manioc	12,638	12,150	11,849	11,849	11,307	10,760	85
black beans	584	453	518	466	490	390	67
wheat	1,104	908	764	1,151	693	1,202	109
corn	1,432	1,597	1,441	1,833	1,735	1,745	122
sugar cane	46,960	49,294	54,749	55,175	59,948	62,867	134
soybeans	1,386	1,750	1,240	1,765	1,565	1,792	129
coffee	839	670	1,108	1,552	1,080	1,462	174

Source: Computed from IBGE (various years).

The political power of the large producers and the urgent need to reduce oil imports diverted PNA away from its social objectives and adversely affected food production. The government opted to direct 90 percent of the available credits and subsidies to cane sugar alcohol, mostly in the Sao Paulo region (Knight and Moran, 1981b: 24), in spite of the obvious merits of manioc. Manioc is more labor intensive, can be grown on marginal land and is less seasonal. Thus manioc alcohol production could have provided year-round employment for the abundant labor force of the Northeast where the soil is adequate for manioc but not for most food crops (Saint, 1982: 225-231).

Because manioc has no political "godfather," its poor peasant growers could not compete with the "sugar sheiks" of the Southeast with their great political power (Erickson, 1981: 167). With the support of government subsidies, the commercial farmers of the Southeast bought out the small farmers and land once devoted to food crops was given over to sugar cane (Knight and Moran, 1981b: 25; Saint, 1982: 229-230).

An import substitution program which could have altered the rural structure became a program which increased the inequity of land holdings and reduced food production. But PNA has significantly reduced the amount of oil imports.

THE Rx WORKS BUT SOME OF THE PATIENTS ARE SICK

The IMF prescriptions for Brazil's debt crisis have kept Brazil solvent but at a price that is unpleasant in the short run and perhaps highly detrimental to Brazilian society in the long run. Carlos von Doellinger of the Economic and Social Research Institute reflects the frustrations of the situation:

> The problem is that despite the fact that we are producing more than we consume and, therefore, generating more wealth per capita, we are forced to remit abroad the major part of the export effort in order to pay the interest on the foreign debt (Quoted in Veja, 3 Oct 1984: 97).

In spite of occasional government statements that the economy is improving, Brazilians perceive a significant decline in real income and purchasing power. One Paulista stated bitterly: "The IMF comes in and harvests all the grain" (Folha de Sao Paulo, 2 Sept 1984: 37).

After the second oil shock, Brazil cut its imports sharply, reduced governmental expenditures, and devalued the cruzeiro to encourage exports in an effort to maintain credit with foreign bankers. With the near collapse of Mexico in 1982, the cautious international bankers insisted that Brazil resort to the IMF for future credit and advice. The austerity measures proposed by the IMF were the type of severe conditions which then U.S. Treasury Secretary Regan had recommended to the IMF (New York Times, 26 Sept 1983: IV, 1). Brazil was to cut government spending, end export subsidies, reduce imports to a bare minimum, and reduce inflation to below an annual rate of 100 percent.

The IMF austerity conditions became more restrictive with the onset of the world recession in 1985 and the continuance of high interest rates, but Brazil has, with a significant amount of grumbling, met most of the requirements. The final price of the debt crisis and the austerity measures is yet to be ascertained, but an initial survey does not lend itself to optimism.

In rural Brazil the emphasis on export agriculture has not only defeated efforts to equalize land tenure but has resulted in increased inequity of land access and has accelerated the growth of an economically marginal rural proleetariat. Export agriculture with modern inputs of machinery, fertilizer and pesticides has pushed peasants off the land. Without land these "boias frias" (cold lunch boys) become low wage, seasonal laborers (Saint, 1981: 98-103). These insecure workers subsist on the margins of rural towns between occasional jobs (Heath, 1981: 270). The problem was exacerbated in the Northeast by the most severe drought in one hundred years. Deaths due to starvation and hunger-related diseases for the 1979-1982 drought are estimated at thirty to sixty thousand (Latin American Regional Report: Brazil, 21 Oct. 1983: 7).

The emphasis on export agriculture and extraction of mineral wealth for quick export may have long range ecological effects. Marginal land is being farmed or mined without regard for soil depletion and erosion. In the Amazon the rain forest is being cleared for mining, farming or ranching in a manner which may result in profound resource impoverishment (Bunker, 1984: 1052). Streams in the Southeast are being polluted by residue from sugarcane alcohol distillers without concern for the aquatic life or the people who use the streams for drinking water (Saint, 1982: 233).

In the cities the IMF (but in reality the debt crisis) is being blamed for unemployment, bread lines, and a loss of national sovereignty. Alan Riding of the New York Times (12 Aug 1984: III, 4) reported that the streets of Rio de Janeiro are filled with itinerant vendors and derelicts. Government health clinics have noted the rapid increase of malnutrition cases, and there is a crime wave brought on by the large number of unemployed youths.

Brazil's last military president, Joao Figueiredo, correctly stated in a December 1984 speech that food production had increased in 1983 and 1984 (O Estado de Sao Paulo, 8 Dec 1984: 2), but government statistics, as shown in the tables above, demonstrate that much of the food was not for consumption by Brazilians. It was, and still is, being exported or turned into alcohol for fuel. Since 1981, the per capita GDP of Brazilians has declined by an average of 3.3 percent per year while the price of food has increased in real terms (O Estado de Sao Paulo, 1 May 1984: 26). Inflation is still at the three figure level and the IMF stipulation that wage adjustments can be no more than 80 percent of inflation means that the buying power of Brazilian consumers will continue to erode.

In addition to the problems of the countryside and the hunger and discontent in the cities, the debt crisis and the IMF prescription present a potentially serious national concern. Brazil has become dangerously dependent on the developed world for the foreign exchange needed to keep the economy functioning. Former Central Bank Director Carlos Viacava reported that 95 percent of the trade surplus needed to service Brazil's foreign debt derives from trade with the ten richest nations and nearly one-half comes from trade with the United States (Visao, 3 Sept 1984: 62-64). Brazil is thus vulnerable to economic fluctuations and to external pressure. Many Brazilians believe the hard line of the IMF was dictated by the U.S. government through the American votes on the IMF board and through Federal Reserve pressure on the American banks to whom Brazil owes large amounts of money (Latin American Regional Report: Brazil, 16 Sept 1983: 7). Brazilians complain that they have lost their national sovereignty and are prisoners of the international bankers and the United States.

The Brazilian government, as of mid-1986, has come to an uneasy truce with the IMF and the lending banks. Brazil is complying with many of the IMF conditions and the interest payments arrive on time so the IMF has chosen to ignore some Brazilian violations in order to avoid a confrontation with the popular civilian government. The government of President Sarney is attempting to reduce domestic discontent by expanding housing, education, and health care programs and by encouraging a shift to food production for domestic consumption (The Economist Intelligence Unit, 1985: 8-12). The problems for the new government are immense, however, because it must satisfy a restive population with food and services (more domestic spending and consumption) and satisfy the IMF bankers by generating surpluses to send abroad for debt service. The two requirements seem to be antithetical in the long run.

The Sarney government has attempted to alleviate the problem in the cities by reinstituting food subsidies and some price controls. In the countryside, the government has increased credits to small farm food producers and is using the 1964 Land Statute to provide some 50 million hectares of land to nearly one and a half million peasant families. This new land reform program was strongly opposed by the landowners and the final version was only 50 percent of the original proposal (New York Times, 13 Oct. 1985: 13). Given the anger of urban residents at their inability to secure adequate food and the anger of the peasants with the loss of access to land, these efforts may be too little and too late.

Some hope does appear, however. Declining interest rates have reduced Brazil's overall debt and thus could lower debt service payments slightly. An expanding world economy, and especially an expanding U.S. economy, could increase Brazil's export opportunities, and since Brazil's exports are tied to the U.S. dollar, the current decline in the dollar vis-à-vis other currencies will make Brazilian exports more competitive. Finally, lower crude oil prices and Brazil's progress toward energy self-sufficiency have reduced the import bill and thus have increased the trade surplus.

These rays of hope may not be enough if the IMF, with U.S. encouragement, continues the hard line. This might be the proper time for the United States to accept the notes of optimism and to push the bankers and the IMF toward a reasonable long-term solution of Brazil's debt problem. Such action would seem to be critical to the survival of Brazil's fledgling democracy. It also would be appropriate for world bankers and political leaders to examine the detrimental effects of policies that push exports at the cost of all else. Export agriculture can create extreme dislocations in the society if it is given a dominant position.

NOTES

1. It must be mentioned that the political power of the latifundists and the large-scale commercial farmers has been more instrumental in stymieing agrarian reform than the debt crisis (Baer, 1983: 327).

REFERENCES

Baer, Werner (1983). *The Brazilian Economy: Growth and Development* (Second Edition). New York: Praeger Publishers.

Banco Central do Brasil (1985). *Boletim*. Rio de Janeiro: Banco Central do Brasil, No. 2.

Berry, Albert and William R. Cline (1979). *Agrarian Structure and Productivity in Developing Countries*. Baltimore: The Johns Hopkins University Press.

Brasil, Fundacao Instituto Brasileiro de Geografia e Estatistica (IBGE). *Anuario Estatistica do Brasil*. Rio de Janeiro: IBGE.

Bunker, Stephen G. (1984). "Modes of Extraction, Unequal Exchange and the Progressive Underdevelopment of an Extreme Periphery: The Brazilian Amazon, 1600-1980." *American Journal of Sociology* 89: 1017-1064.

Economist Intelligence Unit (1985). *Quarterly Economic Review of Brazil*. London: The Economists No. 2.

Erickson, Kenneth Paul (1981). "State Entrepreneurship, Energy Policy and the Public Order in Brazil." In Thomas C. Bruneau and Philippe Faucher (eds.) *Authoritarian Capitalism: Brazil's Contemporary Economic and Political Development*. Boulder: Westview Press: 141-177.

Folha de Sao Paulo. Sao Paulo: Folha Empresa Damanha S.A.

FAO (1985). *The State of Food and Agriculture, 1984.* Rome: FAO.

Furtado, Celso (1984). "Time to Accommodate Debtor Nations." *New York Times* (April 22): III, 3.

Heath, J.R. (1981). "Peasants or Proletarians? Rural Labour in a Brazilian Plantation Economy." *The Journal of Development Studies.* 17: 268-281.

Journal do Brasil. Rio de Janeiro: Journal do Brasil S.A.

Katzman, Martin T. (1977). *Cities and Frontiers in Brazil: Regional Dimensions of Economic Development.* Cambridge: Harvard University Press.

Knight, Peter T. and Ricardo Moran (1981a). *Brazil: Poverty and Basic Needs.* Washington, D.C.: The World Bank.

_____. (1981b). "Bringing the Poor into the Development Process: The Case of Brazil." *Finance and Development* 18: 22-25.

Latin American Regional Reports: Brazil. London: Latin American Newsletters Ltd.

Kutcher, Gary P. and Pasquale L. Scandizzo (1981). *The Agricultural Economy of Northeast Brazil.* Baltimore: The Johns Hopkins University Press.

Mahar, Dennis J. (1979). *Frontier Developmental Policy in Brazil: A Study of Amazonia.* New York: Praeger Publishers.

New York Times. New York: New York Times Company.

O Estado de Sao Paulo. Sao Paulo: O Estado de Sao Paulo S.A.

Pereira, Luis Bresser (1984). *Development and Crisis in Brazil, 1930-1983.* Marcia Van Dyke, Trans. Boulder: Westview Press.

Robock, Stefan (1975). *Brazil: A Study in Development Progress.* Lexington, Mass.: D.C. Heath and Company.

Saint, William S. (1981). "The Wages of Modernization: A Review of the Literature on Temporary Labor Arrangements in Brazilian Agriculture." Latin American Research Review. 16: 91-110.

_____. (1982). "Farming for Energy: Social Options Under Brazil's National Alcohol Programme." World Development 10: 223-238.

Schmink, Marianne (1982). "Land Conflict in Amazonia." American Ethnologist. 92: 341-357.

Smith, Peter Seaborn (1984). "Reaping the Whirlwind: Brazil's Energy Crisis in Historical Perspective." Inter-American Economic Affairs. 37: 3-20.

Syvrud, Donald E. (1974). Foundations of Brazilian Economic Growth. Stanford: Hoover Institution Press.

Veja. Sao Paulo: Editora Abril.

Visao. Sao Paulo: Editora Visao S.A.

World Bank (1982). Brazil: A Review of Agricultural Policies. Washington, D.C.: The World Bank.

4

Agricultural Policy and Food Security in Peru and Ecuador

Cynthia McClintock

INTRODUCTION

In recent decades, while many First and Third World technocrats spoke enthusiastically about "development" for the countries of the South, food production in fact often declined. Hunger is an ever more serious problem in Latin America and Africa. This article analyzes the recent food crisis in Peru and Ecuador. The cases of these two Andean countries are of special interest for various reasons.

First, the policy strategies of the governments in the two countries have been different; by the logic of certain political ideologies, one of the strategies should have succeeded while the other failed. More specifically, in Peru, a major agrarian reform swept large traditional landholders from the countryside and established peasant-run cooperatives on the expropriated holdings. Observers of a progressive political orientation would tend to predict success for such a strategy, while those of a conservative orientation would tend to predict failure.

In contrast, Ecuador's strategy was to maintain private enterprise in agriculture; agrarian reform was limited. Recent Ecuadorian governments have sought to enhance agricultural production primarily by the provision of financial support to this sector. Such a strategy was possible because Ecuador, a member of OPEC, was flush with oil revenues for most of this period. The oil boom was the catalyst for dramatic economic growth in the country; per capita gross domestic product (GDP) increased at an average annual rate of 5.8 percent between 1971 and 1980, second in the Latin American region to Brazil at 5.9percent (IDB, 1985: 152). Ecuador's essentially capitalist, "trickle-down" strategy would typically be favored by political conservatives and criticized by political progressives.

Overall, international analysts have paid much more attention to Peru's agricultural performance than to Ecuador's, and have been largely critical of Peru's record.[1] Only through comparison of the performance of Peruvian agriculture with that of a similar Andean nation can a rigorous assessment be made, however. This article will suggest that, especially considering the greater financial support for Ecuadorian agriculture, the Ecuadorian record is not better than the Peruvian. On various criteria---crop production and employment generation, for example---Ecuador's recent agricultural performance is surpassed by the Peruvian, and on most others it has been about the same. For both countries' citizens, however, the "bottom line" may be the declines in per capita food production, which have been of similar dimensions.

Recently, both Ecuador and Peru have returned to democracy, at least by the conventional North American definition of democracy as electoral procedures for the selection of the administration at regular intervals. Both countries were among the first of the many Latin American countries that have returned to democracy since the late 1970s: Ecuador's transition occurred in 1979 and Peru's in 1980. On the basis of the Ecuadorian and Peruvian experiences, we can thus begin to ask the question: Can the new democratic governments achieve improvements in public welfare when they are confronted by the severest international debt crisis of the twentieth century, and corollary international demands for increases in their agricultural exports and decreases in government expenditure?

While this question cannot be definitively answered for sometime, this article will suggest that, overall, the prospects are not very bright. Peru's new democratically elected president, Alan García the country's second successive democratically elected president) is seeking perhaps more earnestly than any recent Peruvian leader to enhance agricultural production, and some progress seems likely. However, the obstacles are grave. To a certain extent, the problems are topographical: the Andean mountains, where traditionally most food staples are grown in the two countries, are the steepest, most rugged, and highest in altitude in the Latin American region, and they have suffered severe ecological degradation in recent decades (Dandler and Sage, 1985). But perhaps even more important, as the Andean countries have become ever more integrated into the world capitalist economy, the pursuit of private advantage has tended to impair the public welfare. This argument has been advanced for the Latin American region as a whole by Merilee Grindle (1986), Alain de Janvry (1982), and F. LaMond Tullis and W. Ladd Hollist (1986), and the data here for Ecuador and Peru provide further support for this thesis.

Comparison of the agricultural policies of the military governments of the 1970s and the civilian governments of the 1980s is viable in Ecuador and Peru because---in contrast to say Bolivia and Argentina or Brazil and Uruguay---the two countries are rather similar. Per capita income is about the same in the two countries: about $1,055 in Peru and $1,222 in Ecuador as of 1985---among the lowest in South America but higher than most in Central America and the Caribbean (IDB, 1986: 394). Ecuador's population is somewhat more rural than Peru's---about 48 percent to 33 percent (IDB, 1986: 389), but the percentage of the labor force in agriculture is similar, at 40 percent to 45 percent (World Bank, 1981: front matter and CONADE, 1985: 52). A large percentage of both nations' poorer citizens are in agriculture; many agriculturalists are of indigenous Indian stock, living in Andean-highlands peasant communities and following their ancestors' customs.

The two nations' geography is similar: both are Andean nations that include markedly different coastal, highland, and jungle regions. In both, coastal agriculture is the most "modern," highlands agriculture the most "traditional," and the jungle, in particular the lowland slopes between the Andean mountains and the dense jungle, is only gradually being brought under cultivation. Overall, however, the Ecuadorian lands are more fertile; water resources are more abundant on the Ecuadorian coast, and the Andean mountains are not quite so high in Ecuador. Perhaps because the Ecuadorian capital is located in the highlands while the Peruvian is on the coast, transportation and communication between the two regions are somewhat better in Ecuador.

This article will first describe the two countries' agricultural and food record. Then, I will examine recent agricultural policies in the two countries, and seek to understand why these policies were not more successful. I will emphasize the goals and policies of the current era's first two democratic administrations, Fernando Belaerry in Peru between 1980 and 1985, and Jaime Roldós, succeeded after his death in 1981 by Osvaldo Hurtado, between 1979 and 1984 in Ecuador. Special attention will be given to the parcellation of Peru's agrarian cooperatives during this period. I will also discuss the agricultural policies of the current democratic administrations---Alan García in Peru and León Febres Cordero in Ecuador---but more briefly, due to their relatively brief time in power and the inevitable paucity of data on policy outcomes. In the conclusion, the article's arguments will be summarized and some recommendations for agricultural policy offered.

THE RECORD

By most criteria, the agricultural record of both Ecuador and Peru has not been good. Overall trends in gross agricultural production have been positive, but barely so. Food security and food supply per capita have declined slightly in both countries.
While these conclusions are evident, some caution must be exercised in the interpretation of the available data. Cross-national comparison is impeded by somewhat different classification schemes in the two countries. The most important difficulty, however, is the exclusion of coca from almost all data sets and analysis of Peruvian agriculture. Since the late 1970s, coca production has sky-rocketed in Peru (but not in Ecuador); by the mid-1980s, coca had become Peru's most valuable export (Lee, 1985-1986: 145). Because coca production is of course illicit, no official figures about it are collected. But as analysts thus tend to ignore the coca boom in their discussions of Peruvian agriculture, they seriously under estimate Peru's agricultural growth and also underestimate the degree of export orientation in Peruvian agriculture.
At first glance (see Table 4.1), Ecuador's record---3.0 average annual growth in the sector between 1970 and 1982---seems acceptable. Although the trend was less positive than the Latin American average for the sector during this period, and less positive than for Ecuador between 1960 and 1970 (IDB, 1986: 43), it is at least positive. Closer examination of the data in Table 4.1 indicates, however, that in official statistics "agriculture" is defined more broadly in Ecuador than in many countries, and that Ecuador's "agricultural" growth during this period is primarily attributable to jumps in the value of forestry, fishing, and hunting. Non-export crop production in fact stagnated. The performance of agriculture in Ecuador also varied considerably over the decade of the 1970s, and the trend was toward scanter and scanter growth (World Bank, 1984: 5).

Table 4.1
Trends in Value of Agricultural Production: Ecuador
(Average Annual Growth Rates in Percentages)

Ecuador	1970-1982
Total Agricultural Output	3.0
Traditional Export Crops (bananas, coca, and coffee)	1.3
Other Crops	0.4
Animal Production	4.4
Forestry	9.6
Fishing and Hunting	11.9

Source: World Bank (1984: 76).

Peru's overall agricultural performance during the 1970s was no better than Ecuador's (see Table 4.2). The Peruvian government defines "agriculture" as crops and livestock only; fishing, which has been very important in the Peruvian economy, is classified separately. Table 4.2 suggests small increases in livestock production and yet scanter gains in crop production during the decade. Poultry registered the most dramatic increases (Presidency of the Republic, 1982: 15 and Presidency of the Republic, 1984: 236 and 659). As Table 4.2 indicates, exact calculations of Peru's record vary considerably according to the base years for Peruvian agriculture; 1969 and 1970 were excellent years for Peruvian agriculture, whereas 1978-1980 were bad ones (Abusada, 1984: 74).

Table 4.2
Trends in Value of Agricultural Production: Peru
(Average Annual Growth in Percentages)

1970 - 1980	
(in millions of 1970 soles)	
Crops and Livestock	1.3
1969 - 1979	
(in millions of 1970 soles)	
Crops and Livestock	2.0
1969-1979	
(in millions of 1973 soles)	
Crops and Livestock	0.7
Crops	−0.6
Livestock	+3.4

Sources: For figures in 1970 soles, Abusada (1984: 74); for figures in 1973 soles, World Bank (1981: 189).

Table 4.3 shows that, during the first half of the 1980s under democratic governments in the two countries, agricultural performance remained poor. The most dramatic statistics in the table are those for 1983 in both count-countries, indicating the depth of agricultural plummet in the face of extremely adverse weather conditions occasioned by unusual El Niño currents; flooding devastated coastal agriculture in both countries. Peru was also hurt badly by drought in 1980, and again in some highland areas in 1983. Obviously, agriculture in both Peru and Ecuador remain stragically vulnerable to weather conditions. In Ecuador, the positive growth rates for 1984 and 1985 probably did not recoup 1983's losses. Unfortunately, a direct comparison of the Ecuadorian and Peruvian "agricultural" records during the 1979-1985 period is more hazardous than during the 1970s, because by the 1980s the impact of coca was extremely large in Peru.

Table 4.3
Trends in the value of agricultural production under democratic government (percentages)

	ECUADOR		PERU	
Year	Agriculture, Livestock Fishing and Forestry[a]	Crop Agriculture[b]	Agriculture and Livestock	Crop Agriculture
1979	2.9	1.8	3.1[c]	5.6[c]
1980	5.3	5.7	-5.4[c]	-9.9[c]
1981	6.8	4.8	12.7[c]	14.1[c]
1982	2.0	-1.8	3.6[c]	2.6[c]
1983	-13.9	-27.6	-8.0[c]	-13.5[e]
1984	8.8	n.a.	12.9[d]	16.8[e]
1985	4.8	n.a.	2.1	7.8

Sources:

[a] CONADE (1986).
[b] World Bank (1984: 160) and (1984: 76) in thousands 1975 sucres.
[c] Ministry of Agriculture and Food. In millions of soles at 1970 prices.
[d] IDB (1986: 346).
[e] Ministry of Agriculture (1985: 69). In 1979 soles.

In short, gross crop production has essentially stagnated in both countries. Peru's performance on this score is slightly better than Ecuador's: indexing the 1974-1976 period as 100, FAO estimates that gross crop production declined to 97 in Ecuador as of 1984 and rose to 104 in Peru as of the same year (FAO, 1985:81). Moreover, there is some indication that, to a greater extent in Ecuador than in Peru, the record is not worse because of increasing yields for crops sold to agroindustry for livestock rations, such as hard corn and soybeans (CONADE, 1982: 61-62). Table 4.4 shows that, with the exception of rice, the production of all food staples declined in Ecuador between 1970 and 1984. In contrast, Table 4.5 shows that the production of food staples did not plummet as precipitously in Peru (although rice production did not increase as much as in Ecuador, either).

As gross agricultural production stagnated in the two countries, the size of the population jumped and food demand increased. Average annual population growth between 1961 and 1984 in Ecuador has been over 3.0 percent, and in Peru about 2.7 percent (IDB, 1985: 383). These rates were above the region'saverage, at about 2.6 percent (IDB, 1985: 383). Thus, in per capita terms agricultural production actually declined in both countries. According to the FAO (Food and Agricultural Organization of the United Nations), from the three-year 1974-1976 period to 1984, per capita agricultural output fell from the index 100 figure to 89 in Ecuador and 88 in Peru, and per capita crop output fell from the index 100 figure to 73 in Ecuador and 82 in Peru (FAO, 1985: 87-91). The Ecuadorian and Peruvian records were slightly worse than South America's as a whole (FAO,1985: 87-91).

Domestic food supply has also been adversely affected by some agriculturalists' tendency to produce for export, which is especially pronounced in Ecuador. Ecuador's agricultural exports increased from approximately 15 percent of value of agricultural production in 1970, to slightly over 30 percent in 1980-1981, to a startling 50 percent in 1985.[2] In the Peruvian case, the country's traditional agricultural exports, sugar and cotton in particular, were the bulk of Peru's total exports until the mid-1970s, earning Peru about $160 million annually between 1965 and 1970 (FitzGerald, 1979: 81). During the 1970s agrarian reform, however, these enterprises became cooperatives, and they increasingly dedicated their sugar, cotton, and other products to the internal market. Officially, Peru's agricultural exports today are only about $147 million, versus $466 million for Ecuador, a much smaller country (IDB, 1986: 45).

Table 4.4
Production of food staples: Ecuador
(in thousands of metric tons)

Commodity	1970	1972	1975	1980	1982	1984	% Change 1970-1984
potatoes	542	473	489	323	416	390	-28
wheat	81	51	65	31	39	25	-69
rice	96	248	306	381	384	437	355
beans	41	26	n.a.	26	29	26	-37
maize(soft corn)	155	171	90	45	55	57	-63

Sources: For 1970, 1980, and 1984, CONADE (1986: 38); for 1975, IEE and FEPP (1985: 208); for 1972 and 1982, Chiriboga (1985: 40).

Table 4.5
Production of food staples: Peru
(in thousands of metric tons)

Commodity	1970-1971 (average)	1981-1982 (average)	1985[c]	% Change 1970/71-1985
potatoes	1,948	1,747	1,650	-15
wheat	124	110	110	-11
rice	589	744	816	38
beans[a]	49	47	55	12
maize(soft corn)[b]	224	215	230	3

Sources: Unless otherwise indicated, for 1970-1971, Presidency of the Republic (1982: 512) and for 1981-1982, Presidency of the Republic (1984: 236 and 659).

[a] average for 1971-1972--figures from Abusada (1984: 29).
[b] data from Martínez and Tealdo (1985: 6), except for 1985.
[c] data are estimates from Ministry of Agriculture (1985: 122, 124, 130, and 132).

However, this is not the entire story. Peru's boom crop today is coca, which is primarily cultivated on private plots on the eastern lowland Andean slopes. It is estimated that the value of Peru's coca exports was $850 million in 1982, or more than twenty-five percent of the total value of Peru's agricultural output (Lee, 1985-1986: 145; IDB, 1986: 397). These figures dwarf the earlier contributions of sugar and cotton to the economy. If coca were not planted on these lands, and rice and other food crops were instead, food supplies for the country would increase. Of course, coca is very lucrative, and some of the land on which it is planted is not entirely suitable for other crops; still, the coca bonanza, which of course by its clandestine nature does not aid the Peruvian economy as other exports do, has done very little for the bulk of the Peruvian peasantry.

According to the economic principle of comparative advantage, export earnings from agriculture should pay for the imports occasioned by the export emphasis. This logic does not apply very well to Ecuador or Peru today, for many reasons (Hopkins, 1986). Export earnings, especially those from coca, often do not return to the country. Since the early 1980s, with the onset of the international debt crisis, the Latin American countries have desperately been trying to save foreign exchange; thus, while food imports may not be a larger percentage of the import bill than they had been in the 1960s or 1970s, the burden is perceived to be greater. Also, the composition of food imports has changed; whereas in the 1950s or 1960s a larger percentage were non-essential imports, now over half of both Ecuador's and Peru's food imports are cereals and cereal preparations, which provide much of the staple diet of the nation's growing urban populations (IDB, 1986: 44-46). The countries are thus very dependent upon the secure supply of these foods, and may fear the use of food as a foreign-policy weapon by their suppliers. In the case of Peru in the mid-1980s, for example, the García government could fear retaliation by the U.S., its main food supplier, against Peru's policy of strictly limiting debt repayment. Even in normal circumstances, food supply may be uncertain, causing concern among the population. For example, in a poll of Lima citizens in 1986, residents cited food supply as the number one problem in Lima (<u>Debate</u>, Vol, VIII, No. 41, November 1986: 6).

Peru has long been unusually dependent upon food imports. In 1925, for example, food imports were 24 percent of total imports (Thorp and Bertram, 1978: 133); in 1960 and 1962, they were 16 percent of total imports (FitzGerald, 1976: 14; World Bank, 1983: 528). The percentage remained in this range for much of the 1970s and 1980s.[3] In 1985-1986, however, as the Garcia government has sought especially hard to reduce imports, the percentage has risen to as high as 20 percent, although the total value of the imports, about $380 to $400 million, is lower than the roughly $450 million average during 1980-1984.[4] Peru is now spending not only over $100 million for imported wheat, but more than $100 million for rice, dairy products, sugar, and meats (Andean Report, Vol. XIII, No. 9, October 1986: 151).

Ecuador is less dependent upon food imports than Peru. In 1960, food imports were approximately 10 percent of Ecuador's total imports; during the 1970s, as Ecuador earned a great deal of foreign exchange from petroleum, the country's total imports quadrupled, and accordingly food imports declined in percentage terms, to a mere 4 or 5 percent (World Bank, 1984: 130). The total value of Ecuador's food imports has increased, however; in 1970-1971, Ecuador was spending less than $10 million annually in current dollars on food imports; during 1980-1984, the figure was $142 million (World Bank, 1984: 130; IDB, 1986: 45). As oil prices have now plummetted, and Ecuador's capacity to import has been drastically reduced, the nation's food import bill is likely to weigh more heavily on the country.

Neither in Ecuador nor Peru are citizens' food needs met by the countries' domestic supply plus its food imports. Despite Ecuador's high economic growth rate during the 1970s, the country's calorie supply per capita remained extremely low. In 1980, daily calorie supply per capita was only 88 percent of FAO requirements, according to the World Bank (1983, Vol. II: 27), and a mere 76 percent of the requirements established by Ecuadorian institutes (CONADE, 1986). Daily per capita calorie supply in 1978-1980 was a mere 2,092; while this number was about 5 percent more than in the 1969-1971 period, it was one of the lowest in the region---only six calories more than in Bolivia, a country with less than half Ecuador's GDP (FAO, 1982; IDB, 1985: 388). In 1983, with the onset of the international debt crisis, per capita calorie intake fell (Chiriboga, 1985: 61).

Perhaps in part because of the domestic-supply orientation of most of Peru's cooperative and private farmers, and in part because of Peru's more abundant food imports, per capita calorie supply is somewhat greater in Peru than in Ecuador. During the 1978-1980 period, daily per capita supply was 2,166 calories, or roughly 95 percent of FAO requirements (FAO, 1982: 68; World Bank, 1983, Vol. II: 74). However, Peru suffers from extremely unequal regional distribution of foods; for example, in the early 1980s average calorie intake was 96 percent of the FAO requirement in the Lima area, but it was only 92 percent in the central highlands and a mere 72 percent in the northern highlands (World Bank, 1981: 35). According to another study, in the late 1970s daily per capita calorie supply ranged from 2,401 in Tingo Maria city in the high-jungle near coca production centers, to 1,458 in Huancavelica, in the remote, disadvantaged southern highlands (Lajo Lazo, 1986: 77).

Also, Peru's 1982-1985 economic plummet greatly increased poverty in the country. It has been estimated that more than one-third of the Peruvian people were not consuming adequate quantities of food during this period (Andean Report, October 1985: 169). According to Peru's Ministry of Agriculture, per capita food consumption fell from 431 kilograms annually in 1970 to 309 kilograms in 1985 (Caretas, September 1, 1986: 19).

ISSUES OF ACRICULTURAL POLICY: LAND TENURE AND AGRARIAN REFORM

Since the late 1960s, land tenure policies in Ecuador and Peru have contrasted markedly. In Ecuador, where the political power of the landed elite has remained very strong, the role of agrarian reform was primarily to transform large feudal estates into capitalist enterprises (Barsky, 1984; Chiriboga, 1985b). Sometimes referred to as a "junker" agrarian reform, this model has been common in Latin America (de Janvry, 1981). The Peruvian reform was very different: the country's landed elites were swept from power and their capitalist agricultural enterprises were transformed into cooperatives. The reform was initiated and implemented by a leftist military regime, first under Juan Velasco Alvarado and then Francisco Morales Bermúdez; ironically,when democracy was restored in 1980, the new government under Fernando Belaunde Terry sought to revert to more traditional capitalist patterns of land tenure.

Ecuador

Overall, between the early 1960s and 1983, Ecuador's agrarian reform affected about 718,000 hectares, 9 percent of the total land surface, and benefited 78,088 families, 10 percent of all farm families (IDB, 1986: 130). These figures are below the averages for Latin American agrarian reforms (IDB, 1986: 130). Moreover, perhaps a majority of the land adjudicated was property that had previously been worked by the peasant in a feudal arrangement, whereby the peasant had received the right to use the land in return for unpaid work on a contiguous hacienda (Chiriboga, 1985b: 101-102). Generally, these adjudicated lands were of inferior quality (Chiriboga, 1985: 102).

Another major focus of Ecuador's agrarian reform law that tended to encourage capitalism in the countryside was the threatened expropriation of inefficient haciendas. The law prompted many large landowners disinterested in farming to divide and sell their properties. In part as a result, immense holdings are today less numerous in Ecuador, while "medium-size" holdings between twenty and 1,000 hectares are much more numerous and a greater percentage of the total (see Table 4.6 and Barsky, 1984: 355-356).

Table 4.6
Landholding Patterns In Ecuador
(Percent of All Land in Certain Extensions)

	Highlands		Coast	
	1954	1974	1954	1974
Less than Five	11.4	11.9	3.1	4.4
Five to Twenty	9.8	14.2	9.0	12.0
Twenty to One Hundred	14.5	25.8	23.5	31.7
One Hundred to 2,500	35.1	32.9	41.7	40.3
Over 2,500	29.2	15.3	22.7	11.6

Source: Barsky (1984: 355 and 356).

Various Ecuadorian governments aspired to a more radical reform. The moderately leftist military government of General Rodríguez Lara (1972-1976) and the democratically elected Roldós/Hurtado administration (1979-1984) both initially spoke of radical changes on behalf of the poorer peasantry; both were frustrated in their efforts by the political offensives of landed elites. The Rodriguez Lara government at first sharply criticized the problem of land concentration in Ecuador, and considered the establishment of a maximum size for landholdings; however, elite interest groups, especially the chamber of agriculture, protested vehemently in the media and lobbied heavily in the agricultural ministry, and the government's ultimate policy was moderate (Barsky, 1984: 199-272). The Roldós/Hurtado government also hoped to extend the agrarian reform, especially by establishing stricter criteria for farm efficiency; once again, these plans were dashed by intense pressure from the chamber of agriculture, applied on the media and in the factionalized legislature (Handelman, 1984, 1985; Barsky, 1984: 273-294). Ultimately, the swathe of the agrarian reform under both the Rodriguez Lara and Roldos/Hurtado administrations was much narrower than that anticipated in their development plans; in fact, the impact of the agrarian reform has been rather similar across administrations, whether considered "progressive" or "conservative" (CONADE, 1982: 146; Chiriboga Vega, 1983: 11-18).

Peru

The swathe of Peru's agrarian reform was much wider. Between 1968 and 1980, virtually all the large, lucrative haciendas in the country were expropriated (McClintock, 1981). In sharp contrast to Ecuador, where almost two-thirds of private holdings on the coast were over 50 hectares (Barsky, 1984: 356), only about 10 percent of private holdings on Peru's coast were so large in 1980 (Carter, 1986: 4). About 9 million hectares, or 39 percent of Peru's agricultural land, had been adjudicated by 1980, benefiting approximately 431,982 families, or 30 percent of all farm families (IDB, 1986: 130). Again in contrast to Ecuador, the adjudicated lands tended to be the better lands (Matos Mar and Mejía, 1980: 69).

Under the reform, most of the coastal haciendas became production cooperatives, called CAPs, managed by the former hacienda workers; the more lucrative highlands haciendas were transformed into a complex mode of cooperative, called SAIS, while other land was transferred to peasant communities or informal peasant groups (Matos Mar and Mejía, 1980: 67). The biggest winners from the reform were the 120,000-odd ex-hacienda workers who became CAP or SAIS cooperative members; an average cooperative member's property share was valued at about $2,000 in the mid-1970s (McClintock, 1984: 65). Just as haciendas had controlled about half of all Peru's coastal land before 1968, CAPs held about half of it by 1980 (Carter, 1986: 3-4).

Unfortunately, a definitive study of the economic performance of Peru's cooperatives in comparison to private holdings is not available. The evidence to date, however, suggests that production has not varied greatly by enterprise mode. For two coastal CAPs and one highlands SAIS as of the late 1970s, McClintock (1981) found that production increased somewhat in the two cooperatives that had also been performing well as haciendas, and declined slightly in the one that had been performing poorly as a hacienda. For a pooled cross-sectional sample of fifty-nine coastal CAPs between 1975/1976 and 1979/1980, Carter and Alvarez (1986: 5) reported that output growth closely paralleled aggregate agricultural growth over the same period.

Overall, agricultural yields in Peru were very good by Latin American standards both before and after the reform (see Table 4.7 and FAO *Production Yearbooks*). Productivity has been consistently higher in Peru than in Ecuador (see Table 4.7). Only Peru's potato yields have been below Latin American and Ecuadorian averages. If cooperative enterprise had really "decimated the basic structure of Peruvian agriculture," as the U.S. Presidential Agricultural Task Force to Peru (1981: 5) argued, it is dubious that the country's yields for sugar and rice (majorities of which are produced on the cooperatives) could be among the highest in the world some ten years after the reform (FAO *Production Yearbooks* 1980 and 1984).

In various important ways, Peru's cooperatives behaved rather differently from large private agricultural enterprises in Ecuador and other Latin American countries. Table 4.8 points to these variations. Whereas permanent pasture land has increased, encroaching on cropland, in various Latin American countries, such as Mexico, Costa Rica, and Colombia, and to a lesser degree in Ecuador, this trend has not appeared in Peru. In the view of most analysts (Sanderson, 1986; Grindle, 1986), an increase in permanent pasture and livestock in contrast to cropland is deleterious to the public welfare for various reasons: livestock enterprise is land-extensive, requiring little labor, and most of the relatively expensive output is exported or consumed by the middle and upper classes. Domestic food supply for the poor is thus reduced. Table 4.8 shows that, as more land has been devoted to pasture and livestock in Ecuador than in

Table 4.7
Agricultural yields: Ecuador and Peru
(kilograms per hectare)

Commodity	ECUADOR			PERU			LATIN AMERICA (average)		
	1969-1971	1974-1976	1984	1969-1971	1974-1976	1984	1969-1971	1974-1976	1984
sugar	77,689	66,922	75,000	141,926	147,794	136,246	49,005	54,054	63,354
coffee	277	324	273	575	545	640	481	541	579
rice	2,990	2,739	3,133	4,141	4,312	4,720	1,667	1,800	2,134
potatoes	11,818	12,529	11,000	6,413	6,519	8,793	8,389	9,115	11,184

Source: FAO, Production Yearbooks (1980 and 1984).

Table 4.8
Changes in agricultural patterns: Peru and Ecuador

	Arable and Permanent Cropland, 1961/65-1979 (% increase)	Land Under Permanent Pasture 1961/65-1979 (% increase or decrease)	Tractor Use 1969/71-1981 (% increase)	Agricultural Employment 1970-1982 (% increase)
Ecuador	4	16	118	17[a]
Peru	46	-3	31	24

[a] preliminary.

Sources: On land use and tractor use, Grindle (1986: 85, 90). On agricultural employment, for Peru, Ministry of Labor and National Statistical Institute data; for Ecuador, Villalobos (1985: 269) and World Bank (1984: front matter).

Peru, less agricultural employment has been generated in Ecuador. Employment generation is a crucial issue: in 1984-1985, both Ecuador and Peru were facing open unemployment rates of slightly over 10 percent, rates almost double the early 1970s. Underemployment was estimated at about 50 percent in both countries (CONADE, 1986; Andean Report, October 1985: 168).

The difference in employment trends may actually be more dramatic than Table 4.8 suggests. According to recent data from CONADE (1986), the number of agricultural jobs in Ecuador may have dropped in the early 1980s; such a trend is in fact rather common in Latin America today. In Peru, however, various agricultural economists (Scott, 1981; Carter, 1983) have emphasized employment generation as one of the advantages of the cooperative model. Whereas large landowners in capitalist enterprises are rarely interested in employing more workers, cooperative members often want the enterprise to provide full-time and part-time jobs for their relatives and children. Also, while desiring to retain cooperative membership, many members sought to reduce their labor for the enterprise, and especially to hire temporary workers for the more burdensome agricultural tasks.

Further, as noted in the previous section, whereas Ecuadorian agriculturalists became more oriented toward exports in the 1970s and 1980s, Peruvian farmers in most of the country became more oriented toward domestic production. The one important exception to this trend in Peru has been the coca producers. The sugar and cotton that had once been cultivated on the large coastal haciendas for export were offered primarily to the domestic market during the 1970s. The primary reason was not cooperative members' preferences but state regulation. Various laws issued by the Velasco government sought to restrict cooperatives' exports to 50 percent of their production, as well as to encourage the production of food staples. Export earnings for the largest cooperatives, especially the sugar cooperatives, were heavily taxed (Carter and Alvarez, 1986: 21; Rojas Senisse, 1984). The sugar cooperatives have also been discouraged from export during the 1970s and 1980s by low international sugar prices and strict U.S. quotas on sugar (Rojas Senisse, 1984: 63-67). Especially given their relative paucity of international contacts, cooperative members have not been especially aggressive in seeking out new agricultural export opportunities; in turn, international commodity prices have been low, and agricultural import businesses have been skittish about Peru, a relatively remote country with restrictive policies against multinational corporations and considerable political instability.

During the course of the Belaúnde administration (1980-1985), most of Peru's cooperative enterprises gradually dissolved. About 75 percent of coastal CAPS had parcelled by mid-1985, most of them during 1984-1985 (Carter and Alvarez, 1986: 7; Vidal Cobian, 1985: 190). In the parcelling cooperatives, each member received roughly five to ten hectares as a private farm; few arrangements were made for collective use of machinery or other resources.

Why did the cooperatives subdivide? The reasons are various, and include both external pressure upon the cooperatives as well as some members' dissatisfaction with the CAPs. The discussion below is based on interviews with government officials in Lima and Trujillo during 1983-1986, as well as interviews and observation in the two Virú valley cooperatives that I began to study in 1973 (McClintock, 1981). I visited these two cooperatives four times during 1981-1986; one subdivided in 1985, while the other remained a cooperative. I have also drawn on secondary sources, in particular Salcedo (1984), Mendez (1982),and Carter and Alvarez (1986).

At the top government levels, most officials favored private-enterprise systems.[5] President Belaúnde who had espoused an individual, family-farm model for his government's agrarian reform during his first administration in the 1960s, continued to favor such a model in the 1980s.[6] President Belaúnde enjoyed close ties to certain elite families who had owned haciendas; his first minister of agriculture, Nils Ericcson Correa, held considerable coastal land planted in rice.[7] Not surprisingly, most of these families were extremely critical of the military government's agrarian reform, and some hoped that the Belaúnde government would restore their former lands to them. President Belaúnde's ideas were likely to have been strengthened by the Reagan administration's positions; for example, in a letter from President Reagan to Belaúnde about the work of the April 1982 U.S. Presidential Agricultural Task Force to Peru, Reagan writes that the Task Force "emphasizes clearly the need to develop strong private sector and technology responses to urgent food needs" (U.S. Presidential Agricultural Task Force, 1982: 50-51).

Accordingly, although Belaúnde had promised stable land tenure conditions and support for cooperatives during his campaign, he moved in the first months of his administration to encourage private enterprise in agriculture. The administration's November 1980 "Law for the Promotion and

Development of Agriculture" specified in Articles 78 and 80 that the cooperatives would be "re-structured" either by cooperative members' decision or by the Ministry of Agriculture's fiat. Also, enterprises were allowed to use their assets as collateral for loans, and to sell land for the repayment of debts.

While the administration's goal of parcelling out the land was clear, the means for its achievement was not. Some cooperative members desired parcelling out of the land, as I will discuss below; but, they also feared change in a system that had, by and large, worked to their advantage. Especially prior to the onset of Peru's 1977-1985 economic "crisis," cooperative members were benefiting from wage increases, new facilities such as day care centers and housing complexes, and dramatically easier access to education for their children (McClintock, 1981: 226-230 and 288-296). Both organizations representing the Peruvian peasantry----the CNA (Confederacion Nacional Campesina) and the CCP (Confederacion Campesina del Perre vehemently opposed to parcelling out of the coastal cooperatives (Eguren, 1982a; Eguren and Filomeno,1982; Salcedo, 1984). Not surprisingly, therefore, the mere legal change permitting parcellation stimulated only a very small number of cooperatives to subdivide. Even by late 1982, the percentage was certainly below 10 percent of all coastal cooperatives; in the area of my longstanding intensive research, in the Virú valley on the north coast, none had subdivided at that time (Carter and Alvarez, 1986: 6-7; Vidal Cobian, 1985:190).

The external pressures that ultimately pushed most cooperatives to parcelling out the land were primarily economic (Salcedo, 1984; Figallo, 1984: 47). During 1981-1983, as I will document below, Peruvian cooperatives were beset by such adverse terms of trade and such tight credit restrictions that bankruptcy was virtually inevitable; ironically, many of the previously most successful cooperatives, which had enjoyed ample credit during the 1970s, were the hardest hit as they had the largest debts. Bankrupt or nearly so, the cooperative enterprises were then unable to resist pressures of the Agrarian Bank in their area for parcelling out the land.

Did the Belaunde government plan its economic policies for agriculture to achieve the goal of parcelling out the land, or did it just enjoy an unexpected benefit from its overall neoliberal economic policy? The answer is uncertain. Comments from various quarters suggest the possibility that the strategy was planned. In its report, the U.S. Presidential Agricultural Task Force to Peru points out that, under certain conditions, Peru's producer cooperatives "should be allowed to fall into natural bankruptcy. This will permit other, hopefully more talented and aggressive, ownership and management to become involved" (1982: 37). Prime Minister Manuel Ulloa spoke in 1982 of "a governmental decision to stimulate by all possible means the parcelling out of the cooperative enterprises" (Mendez, 1982: 32). Still, ultimately the most intense pressures to parcel out the cooperatives were exerted by regional and local elites; presumably the Belaúnde administration would have preferred more direct control, and more rapid execution, of one of its central policy preferences.

Amid the extreme credit squeeze of 1982-1984, cooperative members became willing to make all kinds of adjustments to gain access to credit; without credit, amid the skyrocketing costs of inputs, they could not plant. In my research enterprises in Virú and elsewhere, most CAP members came to believe that the Agrarian Bank was conditioning loans upon parcelling out the cooperatives (Salcedo, 1984: 36).

Parcelling out the land was not a panacea, however; credit remained very tight. Some cooperative members who became the individual owners of parcels (parceleros) were refused credit on the grounds that they still had to repay the cooperative's debt. Others were refused for lack of clear land titles. At times, parceleros were unable to secure legal titles because they had not carried out a feasibility study for parcelling, or because their feasibility study had not been approved by local officials. The feasibility studies, to be done by expert agronomists, were expensive.

Some of the obstacles facing the cooperative members were mounted by Agrarian Bank and Agriculture Ministry officials who were collaborating with local elites eager to buy rural properties (Caballero, 1984; Salcedo, 1984). Indeed, a few officials may have had their eyes on a property for themselves. Ultimately, in various areas of the country, about 10 percent to 15 percent of former cooperative land may have been purchased by agrindustrialists, ex-hacendados, ex-administrators of cooperatives, government officials, and the like (McClintock, 1985; Ainger, 1983: 14-15).

Finally, however---at least in the Viru valley area of my research---most parceleros did secure enough credit to plant their land. Most often, the parcelero received a loan from another, more prosperous member of the community.

Rates of return were exorbitant; in Viru, after the harvest, most lenders have been receiving their principal back, plus 50 percent of the proceeds. Only two or three wealthy individuals controlled the great bulk of the credit.

In the case of the subdivided Viru cooperatives that I studied, the individual who is most active in making loans to parceleros was formerly a president of the cooperative. While an advocate of the cooperative structure in the 1970s, he was the site's earliest and staunchest advocate of parcelling out land after 1983. Now, he has probably gained more than anyone from parcellings, not only through his usurious loans but also through under-the-table sales of the cooperative's machinery and trucks.

While external pressures were thus in my view by far the most important factor in the dissolution of the cooperatives, it is also true that many cooperative members were not entirely satisfied with the enterprises (McClintock, 1981: 221-235; Gonzales Zuniga 1985: 117). For numerous members, a private parcel was a longstanding dream. Finally, they would be duenos---the lords of their manor, controlling their own destiny. In their view, the cooperative was all right, but not without shortcomings: in particular, a certain level of corruption among some cooperative leaders, and inadequate discipline policies, allowing some lazy members to take advantage of the others (Gonzales Zúñiga 1985: 123).

Will the private parcels perform better on economic and social criteria than the cooperatives did, or worse? This question can only be answered in time. Most analysts see benefits and costs in both systems, and hesitate to propose the imposition of either model on unwilling peasants. Scholarly analysts have perceived corruption and skewed work incentives in the cooperatives just as members did; however, they have tended to think that these problems could be surmounted, especially by new rules linking payment to work effort, as has been done in China and elsewhere (McClintock, 1981: 234-235 and 252-255; Carter and Alvarez, 1985).

Scholarly analysts' major fears about the private-parcel system have been highlighted by Carter and Alvarez (1986). Will important economies of scale be lost? Will tasks that require coordination among parceleros and technical expertise---such as aerial cotton fumigation and pest management continue to be done effectively? Will efficient and just arrangements be made for access to water, roads, and machinery? The social impact is also dubious. Will parceleros recruit family members rather than paid laborers for many tasks---thereby reducing their children's educa-

tional opportunities as well as generating less employment?
What will happen to the schools, day care centers, and
health clinics run by the cooperatives? Will parceleros be
highly vulnerable to market cycles and natural disasters,
and thus perhaps ultimately will many lose their land?

To date, the experience of the subdivided Viru cooperative that I studied has not been particularly promising.
Various former cooperative members have actually fought
physically overland and water rights. Also, whereas the
cooperative had been judiciously diverse in its production,
and production had been oriented for domestic consumption,
in 1986 a great many parceleros were persuaded to plant
asparagus, which is canned and then exported to Europe as a
luxury vegetable (Minaya, 1986). Prices had been high for
asparagus the previous year, and parceleros all hoped for
similar profits in 1986---without considering the economic
implications of an asparagus-production bandwagon. As
production skyrocketed, prices not surprisingly declined and
the factory became more selective in its purchases. Ultimately, the factory gained, but the parceleros did not. In
short, although the cooperatives' agronomists were rarely as
knowledgeable and responsible as they should have been, they
were generally able to impute a greater rationality to production decisions than isolated individuals can.

ISSUES OF AGRICULTURAL POLICY: TERMS OF TRADE

While during the 1970s and early 1980s agriculture
stagnated in both Ecuador and Peru, it did so in Ecuador
despite favorable terms of trade throughout this period. In
contrast, the terms of trade were favorable to agriculture
in Peru only through about 1975, when Velasco was replaced
as president by Morales Bermúdez; after 1975 and especially
during the early 1980s, the terms of trade became unfavorable to Peruvian agriculturalists, especially to cooperative
members. In this discussion of the terms of trade, farmgate
prices, input prices, and agricultural credit are
considered.

Ecuador

Between 1973, with the first increase in oil prices, and
1984, with the inauguration of the conservative president
Febres Cordero, Ecuadorian governments have been rather

generous to agriculture. The corollary question, of course, is why Ecuadorian farmers' response was not more positive.

The Ecuadorian government played a major role in the establishment of farmgate prices for both export crops and agricultural items destined for the domestic market (World Bank, 1984: 78-81). Through extensive import quotas, the government separated its agriculture from the world market. Minimum farmgate prices for items for domestic consumption were guaranteed; they were set primarily on the basis of the production costs, with considerable official concern for price stability (Ruff, 1984; World Bank, 1984: 78-81). The state purchased, stored, and sold selected basic commodities.

Relative to world market prices, the Ecuadorian government established particularly favorable prices for its hard and soft corn and for soybeans (World Bank, 1984: 79). Price protection also appears to have been granted to the potato. Prices for rice were below the world market, but cultivation conditions in Ecuador still allowed attractive profits to rice producers (World Bank, 1984: 80; Ruff, 1984: 11-12). While wheat imports have been subsidized, and have increased substantially in recent years, domestic wheat producers have received minimum guaranteed prices at least equal to the imported wheat (Ruff, 1984: 22-23). Table 4.9 shows that official price increases for such key crops as the potato, rice, wheat and corn tended to be most favorable under the more progressive Rodríguez Lara military regime between 1980 and 1983. Official price increases in 1983 were also occasioned by food shortages during this bad-weather year.

Ecuadorian governments also maintained the availability of key agricultural inputs at relatively inexpensive prices. Water charges were generally below cost (World Bank, 1984: 81). Fuel prices were also low (World Bank, 1984: 81). Although somewhat high by international standards, fertilizer prices were reduced under the Roldos and Hurtado administrations.[8] In Ecuador, all the raw material for fertilizers is imported, and then various fertilizer mixes are manufactured and sold by the public firm "FERTIZA."

During this period, Ecuadorian governments---again, especially the Rodríguez Lara and Roldós/Hurtado administrations---provided increased credit to farmers. Between 1970 and 1982, real agricultural credit from the National Development Bank (BNF) increased about 250 percent (World Bank, 1984: 82). In current dollars, agricultural credit multipled eight times, from about $34 million in 1970 to $292 million in 1981.[9] Whereas in 1970 agricultural credit

Table 4.9
Trend in farmgate prices in Ecuador

Commodity	1973-1975 (sum of percent increases in 1973, '74, & '75)	1976-1978 (sum of percent increases in 1977, '77, & '78)	1980-1982[a] (1982 prices as a percent of 1980)	1982-1983[a] (1983 price as a percent of 1982)
potato chola	81	79	71	193
rice	69	20	22	127
wheat	62	4	28	n.a.
beans (bayo bolon)	48	14	-3	39
soft corn	72	39	43	n.a.
yucca	31	115[d]	38	-2
banana	n.a.	35[d]	42	15
CPI[b]	49	36	25	25 to 30[c]

[a]1982 and 1983 prices are for Quito, not for the nation as whole. 1982 prices are averages for three months (January, June, and December), rather than for the year. 1983 prices are June 18, only.

[b]figures are sums of the CPI figures in the World Bank (1984: 180) which are apparently year-end values and thus not totally comparable to year averages of farmgate prices. CPI data in Ecuador are for lower and middle income groups in three cities.

[c]"1983" figure is for the half-year (to jibe with June price data).

[d]for 1977 and 1978 only.

was only 19 percent of agricultural value added, in 1982 it was 42 percent (World Bank, 1984: 82). Real interest rates were negative: during this period, rates were roughly 6 percent to 9 percent, versus inflation rates in the 10 to 15 percent range.[10] The rate of unrepaid loans has been high (Ruff, 1984: 4).

Commercial banks also provided credit to Ecuadorian agriculture. Indeed, commercial banks supplied more than half the available credit during this period (World Bank, 1984: 82). The interest rates for these loans tended to be high, and positive in real terms.

While Ecuadorian governments favored agriculturalists, they did not favor all agriculturalists equally. Even under Rodriguez Lara and Roldós/Hurtado, who all expressed commitments on behalf of disadvantaged peasants, the trend in credit allocation has been dramatically in favor of larger landowners. Table 4.10 shows that, by 1981, agriculturalists owning less than 10 hectares received less than 1 percent of the government's agricultural credit, whereas those owning more than 500 hectares received 38 percent.

Table 4.10
Patterns of Credit Distribution by Landholding Size: Ecuador (percentage)

	1970	1975	1981
Under 10 hectares	18	3	0.8
10-100 hectares	48	33	30
100-500 hectares	26	32	31
More than 500	9	30	38

Source: IEE and FEPP (1985: 205), from calculations in Banco Nacional de Fomento (1980a and 1981).

In contrast, credit distribution in Ecuador is not distributed particularly unequally among different regions of the country. Guayas, the country's most prosperous coastal area, with approximately 15 percent of the economically active rural population, received 20 percent of agricultural credit in 1981, down from 33 percent in 1970.[11] Disadvantaged highlands provinces---other than Pichincha, where the capital Quito is located---with about 49 percent of the economically active rural population, received 26 percent of official credit in 1981, up from 19 percent in 1970.[12] While the regional distribution of credit is thus somewhat disproportionate to population in Ecuador, we will see below that it is much more disproportionate in Peru.

Peru

Whereas between the early 1970s and the early 1980s, Ecuadorian governments provided considerable terms-of-trade support for agriculture, post-1975 Peruvian governments did not. Between 1975 and 1985, first under the military president Morales Bermudez and then under the elected civilian president Belaunde, Peru was in great economic distress and was repeatedly obliged to accept neoliberal austerity guidelines from the International Monetary Fund. Under these conditions, an active state role in the support of agriculture would have been very difficult.

Table 4.11 shows the dramatic changes in Peruvian farm gateprices in recent years. Between 1971 and 1975, under there formist Velasco government, farmgate prices rose at rates above the consumer price index. This favorable trend was checked, however, under Morales Bermúdez between 1976 and 1979. Considering the trends over the two military regimes, most analysts criticized farmgate prices as too low (World Bank, 1981: 53; Alvarez, 1983: 164; Figueroa, 1984: 112; Billone, Carbonetto, and Martínez, 1982: 92). As of 1980, farmgate prices were about 40 percent below world market prices for rice, 35 percent for beef, 20 percent for sorghum, and 12 percent for corn and cotton (World Bank, 1981: 53). Perhaps most damaging to the poorer highlands peasantry, however, was the government subsidy to imported wheat during this period without a concomitant support price for domestic wheat (Alvarez, 1983; Caballero, 1984). Wheat imports skyrocketed, and thus demand for traditional highlands food staples was discouraged.

Although farmgate prices were considered low in 1980, they declined further in 1981-1982. The dimensions of the plummet are shown in Table 4.11, and are also documented by Abusada (1984: 41-52), Martínez and Tealdo (1985: 1-22); and Gonzales Zúñiga, 1985: 107-108).

Table 4.11
Trends in farmgate prices in Peru

	1971-1975	1976-1979	1980	1981-1982	1983-1984[c]
Increase in Consumer Price Index (CPI)	63[b]	199	59	146	237
Average Increase in On-Farm Prices for Nine Key Agricultural Products[a]	120	236	53	81	259
Potato	168	238	82	29	425
Did On-Farm Prices Increase More or Less Than the CPI?	Much more	More	About the same	Much less	About the same

[a] products are: potatoes, rice, wheat, beans, cotton, sugar cane, hard corn, sorghum, and soya. Data for sugar is not available after 1980; international prices were very low.
[b] see McClintock (1981: 357) for sources.
[c] preliminary data.
Sources: Statistics Office, Ministry of Agriculture, for product prices. CPI data for 1973-1983 from World Bank; for 1984, from Andean Report, February 1984, p.4.

As mentioned previously, conscious political goals may have been behind the fall in farmgate prices. However, the decline was also occasioned by the overall neoliberal economic policies adopted by the Belaúnde government with the encouragement of the IMF and the United States (U.S. Presidential Agricultural Task Force, 1982: 26-28). Accordingly, the state marketing agencies that had been established during the 1970s to regulate prices and marketing for corn, meat, coffee, cotton, wheat, oil, and some other products were dismantled. Agricultural imports were liberalized; Abusada (1984: 38) calculates that tariff and non-tariff barriers against agricultural imports declined by about half between the mid-1970s and the early 1980s. A final important factor, in the context of the open economy, was the drop in world prices for important Peruvian agricultural exports. The government had hoped to stimulate Peru's traditional agricultural exports by gradually eliminating the tax on these items (Abusada, 1984: 41). The impact of this measure paled, however, beside the low world prices for all three of Peru's most important traditional agricultural exports---cotton, coffee, and sugar. In 1983, according to World Bank data, cotton prices were almost 50 percent below 1979 levels, coffee prices about 30 percent below 1979, and real sugar prices were the lowest in fifteen years.

As farmgate product prices declined, input prices rose. Traditionally in Peru as in Ecuador, water charges were below cost (Abusada, 1984: 11). Toward the end of the Belaunde administration, however, water costs were allowed to spiral upwards. In many coastal valleys in 1985, one of the most frequent complaints was high water charges; in the Viru valley, the cost of one hour's irrigation was widely reported to be about five times greater than the average daily wage.

By the 1970s, fertilizers were applied on most Peruvian coastal properties and about 10 percent in the highlands (Martínez and Tealdo, 1982: 104). Under President Velasco, fertilizer prices were low and various measures were taken to facilitate the use of fertilizer. A fertilizer sale monopoly was granted to the state marketing agency ENCI, which subsidized fertilizer prices by 30 to 45 percent, and paid the costs of fertilizer transport to remote regions. Fertilizer sales almost doubled during this period (ENCI, 1977: 20-21). Under Morales Bermúdez, however, fertilizer prices rose markedly, and they did so again under Belaunde. Fertilizer prices increased by about 100 percent more than inflation in 1982, and 20 percent to 50 percent more in

1984-mid-1985 (ENCI, 1985). In 1982 alone, fertilizer sales plummetted by 34 percent (ENCI, 1985).

In comparison to Ecuador, agricultural credit has been scant in Peru. Whereas official credit amounted to over 40 percent of gross agricultural output in Ecuador in 1982, the figure for Peru was a mere 23 percent in 1979, and declined subsequently (World Bank, 1981: 52). In Ecuador, official Agrarian Bank credit was complemented by about equal amounts of commercial bank credit; after 1968, commercial banks withdrew credit to Peruvian agriculture, providing perhaps a mere 5 percent of the total (World Bank, 1981: 52).

Whereas the amount of official agricultural credit jumped about 250 percent in real terms in Ecuador between 1970 and 1982, the trend is only barely positive in Peru during this period. In current dollars, credit in Ecuador multiplied by a factor of eight, versus a factor of four in Peru.[13] Credit became much more available during the Velasco years---almost three times as many current soles were made available in 1975 as in 1970, while inflation over the same period was only about 70 percent.[14] In current dollars, the increase was from $112 million in 1970 to $368 million in 1975, in an era of roughly 3.0 percent inflation in the United States.[15] After 1975, however, the real trend was negative.[16] For example, whereas the amount of credit available in 1975 was $368 million in current dollars, in 1983 the figure was $466 million.[17] As inflation had reached over 7.0 percent in the United States during this period (World Bank, 1983: 257), it can be calculated that, for the same amount of agricultural credit to have been available in real terms, about $625 million should have been provided. Peru thus offered its farmers substantially less real credit in 1983 than it had in 1975.

From agriculturalists' perspectives, the only bright side of post-1975 credit policy was its inexpensiveness. Real interest rates were negative throughout the period (Abusada, 1984: 85, and Agrarian Bank). Under the Belaúnde government, however, the intention surpassed official expectations (Abusada, 1984: 59). Also, the Belaunde government applied much greater pressure to collect interest; indeed, during this period interest rates were often collected prior to the disbursement of the loan (Abusada, 1984: 59).

In accord with the Belaúnde government's hostility toward the cooperative enterprises, credit was diverted away from them. As of 1968, large landholdings---which became cooperatives---received about 53 percent of official credit (Abusada, 1984: 55). Under the Velasco and Morales Bermúdez governments, this percentage increased to over 60 (Agrarian Bank). During the Belaunde years, the percentage declined to below the pre-reform figures: to 48 percent in 1981, 45 percent in 1982, 39 percent in 1983, and a mere 34 percent in 1984 (Agrarian Bank).

Agricultural credit has also been allocated extremely unequally among Peru's geographical regions. While only about 15 percent of Peru's rural population resided on the country's coast in 1980 (Presidency of the Republic, 1981: 468), coastal agriculturalists received over 70 percent of all agricultural credit in the 1960s and 1970s (Martínez and Tealdo, 1982: 102).[18] The percentage declined slightly under the Belaunde government,to about 60-65 percent; but the differential was allocated to Peru's jungle region rather than the poorer highlands area (Agrarian Bank, 1981: 36). The needy highlands area, with about 68 percent of the rural population in 1980 (Presidency of the Republic, 1981: 468), received only roughly 10 to 17 percent of total agricultural credit between the early 1960s and the early 1980s (Martínez and Tealdo, 1982: 102; Agrarian Bank).

Issues of Agricultural Policy: Public Investment

The policy trends just described for agricultural terms of trade in Ecuador and Peru are similar for agricultural public investment. Between the early 1970s and the early 1980s, total agricultural-investment outlays increased considerably in Ecuador, but barely at all in Peru. Moreover, in the opinion of most analysts, the Ecuadorian governments made wiser investments; the Peruvian governments' heavy orientation towards super-high-technology projects, geographically concentrated on the nation's coast, has been sharply critized in many quarters.

Table 4.12 shows the trends in total agricultural-investment expenditures in the two nations between 1972 and the early 1980s. While the figures have been carefully calculated from government documents, they should be considered approximate. Different figures are sometimes reported by different institutions. Also, classifications of the components of public agricultural investment vary, not only between the two countries but to a certain extent within them across time. For example, the Ecuadorian data tends to include more diverse projects---rural roads, for example; in contrast (as we shall discuss below), at least until the early 1980s, the Peruvian data includes dam projects that would ultimately be recognized to be electricity projects to a greater extent than agriculture projects. Perhaps most important, the Ecuadorian data calculations are not consistently reported between the 1970s and the 1980s, and the data for some years are missing. It appears likely that the figures in the table underestimate increases in public agricultural investment in Ecuador. For example, total budget expenditures were reported to be increasing very rapidly, at an annual 38 percent rate in real terms between 1970/71 and 1976/77 (World Bank, 1979: 77).

Table 4.12
Public Investments in Agriculture: Ecuador and Peru

	ECUADOR		PERU	
	in current million sucres	in current million dollars	in current million soles	in current million dollars
1972	262	12	1,306	34
1973	316	13	1,717	45
1974	608	24	4,694	120
1975	1,342	54	7,977	145
1976	1,646	65	8,555	150
1977	458	18	14,544	173
1978	n.a.	n.a.	15,194	97
1979	n.a.	n.a.	24,457	109
1980	2,311*	89*	44,977	156
1981	2,311*	89*	n.a.	209
1982	2,311*	89*	n.a.	203
1983	n.a.	n.a.	n.a.	124
1984	n.a.	n.a.	613,950	156

*Average for the three-year period in current sucres. My calculations from CONADE data. Data are for "fundamental projects" and are not entirely comprehensive.

Sources: World Bank (1979: 395); CONADE (1983); INP (1985); Abusad (1984: 69). Exchange rate data from World Bank (1983).

In absolute terms, Table 4.12 also indicates that total public agricultural investment has been higher in Peru than in Ecuador. Peru's public agricultural investment has also been slightly greater as a share of all public investment. The share rose from under 10 percent in the 1960s and early 1970s in Peru to 14 to 17 percent for the remainder of the 1970s, and then dipped again in the early 1980s to 10 or 11 percent (Portocarrero, 1982: 440; INP, 1985; Central Reserve Bank, 1984). In contrast, agricultural allocations peaked as a share of all public investment in Ecuador in 1975 at about 13 percent; as of 1970, the percentage share had been merely about 7 percent (JUNAPLA, 1978). Only during the 1980-1982 period were the Ecuadorian and Peruvian shares similar, with an 11 percent share in Ecuador at this time (CONADE, 1983). In both countries, but especially Peru, the petroleum sector was absorbing the largest percentage of total investment during the mid and late 1970s---about one-quarter of total capital outlays in Peru and one-fifth in Ecuador (Portocarrero, 1982: 440; JUNAPLA, 1978). These figures should be interpreted in the context of the relative size of Peruvian and Ecuadorian agricultural GDP; the Peruvian is about double the Ecuadorian (CONADE, 1968; World Bank, 1981: front matter).

Perhaps more important than the total amount of money provided for agricultural investment, however, is the wisdom of the projects selected for funding. Unfortunately, overall, agronomists have been very dubious about both the cost-effectiveness and the social benefits of the Peruvian investments, which have heavily emphasized super-scale irrigation projects. By the late 1970s, virtually every analysis---from the right, the left, or wherever, criticized Peru's super-scale irrigation projects (Agency for International Development, 1982; World Bank, 1979, 1981, 1983; Guerra Tovar, 1986; Urban, 1986; Portocarrero, 1982 and 1983; Eguren, 1980; Acosta et. al, 1982). These projects are spectacular engineering and construction schemes that, in some cases, are to transport water thousands of miles from the high Andean mountains to the coast; most of the projects are aimed to generate electricity as well as open new lands for cultivation.

The costs of the Peruvian projects, especially of Majes in the southern part of the country, are astronomical. Abusada (1984: 65) indicates that the total cost for the four largest Peruvian irrigation projects will be over $2 billion, and for Majes alone over $1 billion. In contrast, the World Bank (1979: 377) estimated a few years earlier that the total costs in 1979 dollars for three of the largest Ecuadorian irrigation projects identified in Table 4.13 would be "only" about $300 million. As costs have invariably been underestimated (Guerra Tovar, 1986), these figures are probably low. The World Bank has repeatedly identified these projects as cost-ineffective (1979: 343; 1981: 49-50; 1983: 49-50). Table 4.13 contrasts the costs for "large" irrigation projects to smaller-scale alternatives in both Peru and Ecuador. The costs, especially for Majes, are very high incomparison to the current market value of one hectare of similar coastal lands which was only about $1,500 to $3,000 in 1982 (Agency for International Development, 1982: 15). Also, as the new lands are expected to be of good quality, there has been a great deal of land speculation, and probably only well-to-dc families will become settlers.

Super-scale irrigation projects have been criticized on further grounds (Urban, 1986; Abusada, 1984). As very advanced technology is required, the governments must contract international firms and import a great deal of the material, and so many of the immediate benefits from the projects flow abroad. Also, the construction periods for these projects are very long; for Majes, the period may be twenty-three years. Over such a relatively long time, new complications are almost inevitable, and meanwhile there are few domestic beneficiaries. Some analysts have also pointed out that, as these projects are experimental ones in rather atypical ecological zones, environmental problems have often emerged.

Analyzing the agricultural and social needs of Peru and Ecuador, agronomists have recommended alternative projects to the countries' governments (World Bank, 1979: 342; World Bank, 1981: 49-50; Urban, 1986; Dandler and Sage, 1986). Analysts have submitted that the development of new coastal lands at exorbitant costs is nonsensical when at least 40 percent of the land already available for agriculture is not cultivated in Peru and Ecuador (Dourojeanni Ricardi, n.d.: 14; World Bank, 1983: 85). As a result of soil erosion on the steep Andean mountain sides, as well as salinity and water logging on coastal lands, previously fertile land has been deteriorating. Posner and McPherson (1982: 349) believe that as much as 87 percent of Peru's land loses more than ten tons of top soil per hectare annually.

Table 4.13
Peruvian and Ecuadorian Irrigation Projects: Relative Costs

	Cost per hectare (Thousands of U.S.$)[a]	Cost per family benefitted (Thousands of U.S.$)[a]
Peru		
Large Projects		
Majes	27.3	163.7
Chira-Piura	3.3	16.5
Jequetepeque-Zaña	3.4	9.4
Tinajones	2.6	10.1
Small and Medium Projects		
Rehatic	1.8	8.6
Meris	1.2	1.3
Ecuador		
Large Projects		
Carrizal-Chone	5.6	n.a.
Tahuín	3.9	n.a.
Jubones	1.3	n.a.
Daule-Peripa	1.2	n.a.
Small Projects		
Integrated Rural Development	0.1	n.a.

[a] In 1981 dollars for Peru and 1979 dollars for Ecuador.

Sources: For Peru, Presidencia de la República (1981: 298). For Ecuador, CONADE (1979).

To correct these problems, agronomists have suggested reforestation programs, new drainage networks, desalinization efforts, and small-scale irrigation initiatives. Not only are these projects relatively inexpensive, but they employ technologies and materials largely available domestically, and tend to benefit primarily poorer peasants within a relatively short time. These recommendations have been adopted to at least a certain extent in both Ecuador and Peru. Table 4.13 identifies programs that reflect agronomists' concerns: Rehatic (for coastal land rehabilitation) and Meris (small-scale irrigation) in Peru and Integrated Rural Development Projects in Ecuador.

Tables 4.14 and 4.15 contrast the Ecuadorian and Peruvian governments' agricultural investments in the 1970s and 1980s. A comparison of the data in the two tables show that Peruvian governments have committed much larger percentages of their investment budgets to super-scale irrigation projects than Ecuadorian governments. The share in Peru has been well over 50 percent under all recent governments, and in some years even over 90 percent (Guerra Tovar, 1986: 158) versus under 25 percent in Ecuador. Moreover, as we saw in Table 4.13, Peru's "large" irrigation projects, especially Majes, are "larger" and more cost-ineffective than those of Ecuador's. Also the Roldós/Hurtado government deferred expenditure for the most expensive project, Carrizal-Chone, citing legal problems. Under this government, the best-funded "large," "high-technology" project was Daule-Peripa, which as Table 4.13 indicated was the least expensive, costing only about as much as many Peruvian projects of the smaller scale.

The Roldós/Hurtado government increased Ecuador's agricultural programs for the poorer peasantry somewhat more than Table 4.14 specifies. During the 1980s, the government's effort to promote integrated rural development was considerably more concerted than in the past. A new institution (SEDRI, or Secretare Desarrollo Rural Integral) was created specificallyto advance the program; the number of projects increased, and during 1980-1982 they advanced towards completion more rapidly than most among the fourteen "fundamental" rural development projects (CONADE, 1983). On the other hand, the actual initiatives were relatively similar to those under the "rural development" program of the 1970s efforts. Also, the 1980s integrated rural develpment programs were criticized on various counts. Some analysts, especially those on the left, considered the program a palliative for the grave social injustices in the countryside that could only really be ameliorated through agrarian reform; they argued also that peasant community elites would be the major beneficiaries from the program. Other analysts, particularly those on the right, criticized the amount of partisan politicking injected into the program by SEDRI officials.

Further, as the data in Table 4.14 for the 1980s encompass only "fundamental projects," the table does not

Table 4.14
Trends in agricultural investment in Ecuador

	Percentage of Total Agricultural Investment	Average Annual Investment[a]	Percentage of Total Agricultural Investment	Average Annual Investment[a]
I. Large Irrigation Projects	16	7.3	22	19.8
Carrizal-Chone		3.4		2.2
Jubones		3.8		0.7
Daule-Peripa		0.1		13.5
Tahuin		0.0		3.4
II. Small and Medium Irrigation Projects	13	6.1	n.a.	n.a.
III. Rural Development and Integrated Rural Development	16	7.3	21	18.9
IV. Other (Agrarian Reform, Colonization, Agriculture and Livestock)	55	25.4	57	50.3

[a] in millions of dollars.

Sources: JUNAPLA (1977: 1980-1981) and (1978); CONADE (1983). 1980-1982 data are for top-14 projects. Calculaions for 1970s in current sucres; for 1980-1982 in 1979 sucres. "Top-14" priority projects in agriculture include national parks and fishing.

Table 4.15
Trends in agricultural investment in Peru

	1970-1976		1978-1979		1981-1983[b]	
	Percentage of Total Agricultural Investment	Average Annual Allocation (Mil. of $)[a]	Percentage of Total Agricultural Investment	Average Annual Allocation (Mil. of $)[a]	Percentage of Total Agricultural Investment	Average Annual Allocation (Mil. of $)[a]
Large Irrigation Projects	72[c]	86	64 to 85[d]	n.a.	67	122
Majes	n.a.	40	60[h]	54[e]	23	42
Chira-Piura	n.a.	38	n.a.	n.a.	26	47
Jequetepeque-Zana	n.a.	n.a.	n.a.	n.a.	10	19
Tinajones	n.a.	6	n.a.	n.a.	5	10
Small and Medium Irrigation Projects (Meris, Rehatic, etc.) 4.5[f]	6	69[g]	n.a.	10	18	
Other (Development Corporations, Colonization)	n.a.	n.a.	n.a.	n.a.	23	--

Sources:
[a]1979 World Bank Study, for 1971-1976 only; Urban (1986: 196).
[b]Abusada (1984: 69). 1983 figures are preliminary and overestimate expenditure on small-scale projects.
[c]Portocarrero (1982: 444).
[d]Eguren (1980: 41) suggests the lower figure; Portocarrero (1982: 441) the higher. See also Urban (1986: 196).
[e]Central Reserve Bank data.
[f]estimate for 1970-1975 from Ministerio de Economia, Finanzas, y Comercio data.
[g]estimate from data for 1975-1978 plan in Institute Nacional de Planificacion (1975: 87-88).
[h]World Bank (1981: 50), for 1979 only.

include all relevant 1980s rural programs. In particular, it excludes the program "Foderuma," or Development Fund for the Marginal Rural Sector, which was established under the auspices of the Central Bank. To an even greater extent than Integrated Rural Development program, Foderuma aimed to benefit poorer communities and sought to foster social and political organizations in the communities; also, its projects were more numerous than those of SEDRI. However, Foderuma did not enjoy as ample resources as the integrated rural development program. From mid-1978 to early 1983, Foderuma invested 580 million current sucres in the countryside (roughly $20 million), about one-quarter of SEDRI's allocations (Banco Central de Ecuador, 1983).

In contrast to Ecuador, successive governments in Peru lopsidedly favored super-scale irrigation projects. Through the 1970s and early 1980s, they allocated about $100 million annually to super-scale irrigation projects. The largest and most cost-ineffective project, Majes, which has been strongly criticized by both international and domestic economists since the late 1970s, received a staggering one-quarter to one-half of these funds (see Table 4.15). While the Belaúnde government did gradually reduce its commitment to Majes, the project still received $281 million during its five-year term, making it the government's sixth-highest investment priority (Central Reserve Bank, 1986). In contrast, small-scale irrigation projects ranked number 32 on this list with a total $51 million over five years and land rehabilitation number 40 with $40 million (Central Reserve Bank,1986).

The Belaúnde government's primary explanation for this pattern of expenditure was that, once large sums had been invested in Majes and similar projects, they should not be stopped; contracts had been signed and expectations raised, and the funds already spent would be wasted if new money was not committed. Speaking confidentially, however, most economists doubted these arguments. At a minimum, economists usually arguedthat the expenditures should be deferred until Peru was out of its grave economic crisis. There has also been a broad consensus that these super-scale dams were pork-barrel projects, continued for officials' personal gain.

Although the available data are not comprehensive, it appears that Ecuadorian public investment has been more equitably distributed among the country's geographical regions than the Peruvian. During the 1970s, more than half of Ecuadorian investment that was targeted at a specific department (excluding multidepartmental investments) was targeted for the relatively poor highlands region (JUNAPLA, n.d.: 129). In the early 1980s, about two-thirds of all agricultural projects were targeted for highlands departments (Ministry of Agriculture and Livestock, 1983). In contrast, Peruvian public investment has been very biased towards the more prosperous coastal region and against the poorer highlands area. Wilson and Wise (1986: 98) calculate that 61 percent of all Peruvian public investment between 1968 and 1980 was destined for coastal departments, versus a mere 10.5 percent for highlands departments (and 16 percent multidepartmental). The percentage for the highlands increased under the Belaunde government; the figure was 28 percent during 1981-1982 (Wilson and Wise, 1986: 98), with roughly 5 percent to 18 percent multidepartmental investment and remained at this level through 1985 (Instituto Nacional de Planificacion, 1985). However, especially with the Belaúnde government's emphasis on electricity programs, a considerable share of the funds spent in a highlands region were for hydroelectric projects that were supposed to benefit coastal residents (Central Reserve Bank, 1986).

Agricultural Policies Under the Second Democratically Elected Administrations

A second consecutive constitutional succession occurred in Ecuador in August 1984, and in Peru in July 1985. The ideological position of both new democratically elected governments was very different from their predecessors'. While it is difficult to pinpoint any government's exact position on an ideological spectrum, a consensus view would be that the center-left Hurtado administration was succeeded by the rightist Febres Cordero government; and, in the case of Peru, the center-right Belaúnde was succeeded by the leftist Alan García. Both new governments initiated dramatic changes in their countries' agricultural policies. Unfortunately, these policies are still too recent to be fully documented, and the results of the policies will not be clear for some time. This section will thus provide only a brief overview of the policies in the two nations.

The Febres Cordero government sought to advance free-market principles for agriculture.[19] Agrarian reform was not on the government's agenda. The integrated rural development programs that had been a priority for the Roldós/Hurtado administration were downplayed; SEDRI, the institution responsible for the program, was transferred among several ministries and morale was reported to be very low. Just as a U.S. Presidential Agricultural Mission had visited Peru in 1982, a similar mission went to Ecuador in 1984, and recommended the implementation of a host of free-market measures: the end of minimum guaranteed prices for farmers; more secure loan collateral provisions for agricultural lenders; devaluation, in part to encourage agricultural exports; and, overall, an enhanced role for the private sector in agriculture. These recommendations were adopted by the Febres Cordero government. Although data are not yet available, public investment in agriculture may have declined.

The government's policies have not hurt the agricultural sector in the aggregate. Indeed, perhaps in part because of the government's free-market policy, large agricultural loans were made in 1985-1986 to Ecuador by the Inter-American Development Bank and the World Bank, and credit has thus accordingly become rather abundant (Business Latin America, May 26, 1986: 156). Overall growth in the Ecuadorian agricultural sector, reported to be 4.8 percent by CONADE in 1985, promised to be as good or better in 1986 (Business Latin America, May 26, 1986: 156).

However, the distribution of these benefits has been highly skewed among Ecuador's farmers. It will be recalled that, increasingly, Ecuador's agricultural credit has been destined for large landholders. Moreover, the export bias in Ecuadorian agriculture, already strong, has become even more pronounced (Business Latin America, May 26, 1986: 156). The increase in Ecuador's agricultural production in 1986 was primarily due to jumps in the value of fish, coffee, and banana exports; the production of non-export crops was still 19 percent lower in 1986 than in 1981 (Latin American Regional Reports---Andean Group, 29 January 1987: 7, and Latin American Economic Report, 31 December 1986: 6). Ecuador's peasant organizations are restive, decrying "land famine" (Latin American Regional Reports---Andean Group, 29 January 1987: 7).

In contrast, the new García government disdained free-market principles, preferring immediate measures to enhance social welfare, especially for the rural poor. Plans were announced for a restructuring of the highlands SAIS cooperatives, particularly in the Puno region, in order to provide more land to impoverished peasant communities. Agrarian Bank credit for agriculture jumped by almost 50 percent in real terms (Andean Report, January 1987: 4). Interest rates were reduced; the maximum rate became 55 percent on the coast, 20 percent for most of the highlands, and zero for the "Andean Trapezoid"---the new government's term for the five destitute southern highlands departments Ayacucho, Apurimac, Huancavelica, Cuzco, and Puno, plus highland areas in Arequipa, Moquegua, and Tacna (Martínez, 1985: 35; Central Reserve Bank, 1986). The prices for most key agricultural inputs were slashed; accordingly, sales of fertilizer tripled between 1985 and 1986 (Andean Report, January 1987: 3). Farmers were also guaranteed reasonable prices for basic agricultural products by the government; the total cost of the government's subsidies, which were paid primarily for rice, corn, and sugar, has been estimated at about $110 million (Andean Report, January 1987: 5).

Public investment in agriculture increased just slightly more than inflation (Instituto Nacional de Estadística, 1986). Although comprehensive data are not available, it appears that a greater effort was made to assure that more investments would benefit the rural poor. Public investment in the Andean Trapezoid was increased; $120 to $160 million was allocated to this region in 1986, and about $30 to $35 million for Ayacucho (Central Reserve Bank, 1986). Spending on three super-scale irrigation projects---Majes, Chira-Piura, and Jequetepeque-Zaña--declined from about $55 million in 1985 to $45 million in 1986.[20]

In García's 1985 presidential campaign, one promise was made to agriculturalists that was not kept: food imports were not decreased, and Peru actually became more dependent on food imports in 1986. For 1986, it was estimated that 51 kilograms of food were imported per capita---versus 34 in 1985 (Caretas, September 1, 1986: 19). As previously, the prices for imported wheat and pasta were not raised to levels that would persuade Peruvian consumers to purchase larger quantities of domestic variations of these food staples. However, the García government benefited from very low international grain prices, and was able to sell some of the imports for a profit to local millers and processors, while still retaining low food costs; the profit was given to a fund for subsidies on various Peruvian crops (Andean Report, January 1987: 5-6).

The results of the García government's policies appear to have been very favorable. Agricultural production was up 3.6 percent in 1986; the biggest jumps were for the food staple potatoes, and also for chicken (Andean Report, January 1987: 2-5). Real farm income rose by as much as 25 percent (Andean Report, January 1987: 1). Although relatively well-to-do farmers gained along with smaller landowners, the problem was not as severe as in Ecuador, as Peru's largest landowners left the countryside amid the agrarian reform. Also, it was estimated that, for various reasons, including the government's temporary employment program as well as its agricultural policies, wages for highlands farm laborers doubled in real terms between mid-1985 and late 1986 (<u>Andean Report</u>, January 1987: 3).

CONCLUSION

As this article has analyzed two countries where neither agricultural growth nor food security have been enhanced much, if at all, in recent years, it can only offer suggestions about which agricultural policies will not succeed, rather than about what policies will succeed. Evidently, neither a policy based almost exclusively on agrarian reform, without favorable terms of trade or agricultural investment---as in Peru---nor a policy based primarily on economic support for agriculture, without a significant agrarian reform---as in Ecuador---is sufficient. Perhaps a combination of the two models is necessary: agrarian reform and economic support from the government. However, to assure agricultural growth with equity in the Andean nations, considerably more thought seems necessary on many scores---for example, about whether cooperatives or private parcels should be encouraged, and about which agricultural products should be promoted, and how.

The comparison of agricultural policies and agricultural performance in Ecuador and Peru across various types of governments has illuminated various points. It was evident, for example, that Peru's public investment has been much more skewed than Ecuador's towards exorbitant super-scale hydroelectric projects, rather than toward smaller-scale irrigation and land-rehabilitation programs that would be more cost-effective and more advantageous for the poorer highland peasantry. The unusual extent to which Peruvian public investment has been biased in favor of the coast against the highlands also became clear.

Perhaps most important, while Peru's agricultural performance has been at best fair over the last fifteen years, Peru's record seems less deficient in comparison to Ecuador's. Many critics have blamed Peru's low agricultural growth rates on the country's 1970s agrarian reform. This argument, which has scant empirical foundation in Peru in any case, becomes even more difficult to sustain given that the record of Ecuadorian agriculture, with respect to either growth or food security, has barely---if at all-surpassed the Peruvian. Indeed, if coca were included in the official statistics, Peru's agricultural growth is likely to have exceeded Ecuador's considerably.

Scholarly analysis of Latin America should incorporate analysis of the coca agroindustry. The question about private gain versus public welfare that has been very important to recent scholarly research on Latin American agriculture can only be addressed for Peru if much more information becomes available about the coca agroindustry. The picture of Peruvian agriculture that emerges from the official data is of an almost entirely domestically-oriented agriculture, and this picture is of course grossly inaccurate.

Has the establishment of democratic government brought about new policies that are more favorable to agriculture and to rural popular groups? The evidence in this article suggests not. For whatever reason, many democratically elected governments, such as Belaúnde in Peru and Febres Cordero's in Ecuador, may not bee specially sympathetic to the goal of growth with equity in agriculture. However, other democratic governments, such as García in Peru, have been committed to this goal. It is to be hoped that, from the various new policy initiatives under the Garcia government, more answers will emerge to the questions about viable strategies for the achievement of agricultural growth with equity in the Andean context.

NOTES

*I would like to thank the Graduate School of Arts and Sciences at George Washington University and the School of Public and International Affairs at George Washington University for their support of my summer research on this topic. I am also grateful to Luis Deustua and Jose Garcia for their help in the collection of the data, and to Manuel Chiriboga and Jaime Quintana for comments on previous drafts.

1. See, for example, World Bank (1981: 48-54); Alvarez (1983); AID (1982); and stories by Juan de Onis for The New York Times during the late 1970s.

2. Sources for agricultural export data are CONADE (1982: 85) and CONADE (1986). Source for total value of agricultural production is IDB (1986: 397).

3. Calculations from Lajo Lazo (1986: 97); IDB (1986: 397); World Bank (1983: 528).

4. Value of agricultural imports estimated in *Financial Times*, Sept. 26, 1986: 4; and by Agriculture Minister in interview in *Vision*, July 27, 1986. Total import estimate from IDB (1986: 396) and *Andean Report*, October 1986, Vol. XIII, No. 9 (October 1986): 151.

5. Ministry of Agriculture and Agrarian Bank officials in Lima were very open on this score in my interviews with them in January-February 1983. See also Caballero (1984: 46) on the views of the Belaunde administration's agriculture ministers.

6. Interview with President Belaúnde, June 9, 1981; see also Carter and Alvarez (1986: 4).

7. See various issues of *AgroNoticias*; Gilbert (1977: 151-152); Jaquette (1971: 139-143). While the Belaunde family was not generally considered part of Peru's oligarchy, it was one of the "most distinguished" families in Arequipa's "traditional upper class" (Astiz: 2, 1969: 112).

8. With the period May 1978-April 1979 as a baseline, calculations from official data show that the cost of the three most important fertilizer mixes increased about 58 percent by January 1983, considerably less than the consumer price index overall, which rose about 75 percent. Consumer price index data are from INCE (1982: 12-13), where May 1978-April 1978 is calculated as the baseline. See also World Bank (1984: 81).

9. Calculations from data in Banco Nacional de Fomento (1980a: Tables 27-30), Banco Nacional de Fomento (1980b) and (1981: 20). For 1982, Banco Nacional de Fomento, Boletín Economica No. 25 (April 1983): 4 (provisional data). Includes credit to agriculturalists for industrial, handicraft, commerce, and livestock purposes. Exchange rate data calculations based on CONADE (1986).

10. For interest rates, see resolutions by the Junta Monetaria published by the Central Bank; for inflation rates, CONADE (1982) and World Bank (1984).

11. Population data from CONADE (1982: 14). Credit data from Banco Nacional de Fomento (1980a: Table 44) and Banco Nacional de Fomento (1980b and 1981: Annex 20).

12. Idem.

13. Credit data from the annual Memorías of the Agrarian Bank. Inflation and exchange rate data from McClintock (1981: 357) and World Bank (1983: 144-145).

14. Idem.

15. Calculation from exchange rates in World Bank (1983: 144-145); United States inflation from World Bank (1983: 256-257).

16. Data on credit from annual Memorías of Agrarian Bank. Calculations in real terms made through inflation data and currency exchange data from the World Bank.

17. Idem.

18. Unfortunately, local offices of Peru's Agrarian Bank do not correspond to official "department" (province) boundaries; this calculation is based on amounts loaned by bank offices in predominantly coastal or predominantly highland areas. It is possible that some of the loans are allocated to farmers in other areas. While these figures should thus be considered approximate, there is a very broad consensus about the unequal geographical distribution of credit in Peru (World Bank, 191: 52).

19. Information in this paragraph is drawn primarily from an interview with Manuel Chiriboga, July 3, 1986.

20. Calculations from Central Reserve Bank (1986) and Ministerio de Economía y Finanzas (1986).

REFERENCES

Abusada, Roberto (1984). Politica Agraria en el Perú 1980-1983. Lima: Report prepared for USAID (March).

Acosta, A., et. al. (1982). Ecuador: El Mito del Desarrollo. Quito: Editorial El Conejo.

Agrarian Bank (of Peru) (1970-84). "Memoria Anual." Annual publication. Lima.

AID (Agency for International Development) (1982). Report of the United States Presidential Agricultural Task Force Peru. Washington, D.C. (April).

Ainger, Hilary M. (1983). "Peasant Households, Production Cooperatives and Agricultural Policy 1976-1981: A Peruvian Case Study." Paper presented at the Latin American Studies Association meeting, Mexico City.

Alvarez, Elena (1983). Política Económica y Agricultura en al Perú, 1969-1979. Lima: Institut de Estudios Peruanos.

Argones, Nelson (1985). El Juego del Poder: De Rodígues Lara a Febres Cordero. Quito: Corporacion Editora Nacional.

Astiz, Carlos A. (1969). Pressure Groups and Power Elites in Peruvian Politics. Ithaca: Cornell University Press.

Banco Nacional de Fomento (1980a). "Boletin Estadístico 1970-1979." Quito.

_____ (1980b). "Informe de Labores 1980." Quito.

_____ (1981). "Informe de Labores 1981." Quito.

Barsky, Osvaldo (1984). La Reforma Agraria Ecuatoriana. Quito: Corporacion Editora Nacional and FLACSO.

Bates, Robert H. (1981). Markets and States in Tropical Africa: The Political Basis of Agricultural Policies. Berkeley: University of California Press.

Billone, Jorge, Daniel Carbonetto, and Daniel Martínez (1982). Terminos de Intercambio Ciudad-Campo 1970-1980: Precios y Excedente Agrario. Lima: CEDEP.

Caballero, José Maria (1984). "Agriculture and the Peasantry under Industrialization Pressures: Lessons from the Peruvian Experience," Latin American Research Review, Vol. XIX, No. 2.

Caballero, Victor (1984). La Crisis Agraria en el Perú. Lima: Instituto de Apoyo Agrario.

Carter, Michael R. (1983). "Revisionist Lessons from the Peruvian Experience with Cooperative Agricultural Production," Department of Economics, Georgetown University.

_____ (1985). "Cooperativas, Parcelacíon y Productividad: por una Alternativa Mixta," Socialismo y Participacíon, No. 29: 45-51.

Carter, Michael R. and Elena Alvarez (1986). "Changing Paths: The Decollectivization of Agrarian Reform Agriculture in Coastal Peru." Paper forthcoming in a volume to be edited by William C. Thiesenhusen. Department of Agricultural Economics, University of Wisconsin at Madison.

Central Reserve Bank (of Peru) (1984). "Inversíon del Sector Publico, 1980-1984." Lima, February 27, 1985, mimeo.

_____ (1986). "Ayuda Memoría." Lima: mimeo.

Chiriboga Vega, Manuel (1983). "El Estado y las Políticas Agrarias en Ecuador, 1964-1982." Quito: FLACSO (unpublished manuscript).

_____ (1985a). "El Sistema Alimentario Ecuatoriano: Situacíon y Perspectivas," Ecuador Debate: La Cuestion Alimentaria (September): 35-84.

_____ (1985b). "La Crisis Agraria en el Ecuador: Tendencias y Contradicciones del Proceso Reciente," in Lefeber (1985): 91-132.

CONADE (1979). Perfiles de los 41 Proyectos de Inversí Fundamentales. Quito.

_____ (1982). Indicadores Socio-Económicos. Quito.

_____ (1982b). "Appreciación de la Política los Programs de Desarrollo Social en el Año 1981." Quito.

_____ (1983). "Resumen de las Inversiones y Avance Programdos y Realizados en los Proyectos Fundamentales hasta 1982." Quito.

_____ (1985). Plan Nacional de Desarrollo 1985-1988. Quito.

_____ (1986). "Ecuador: Indicadores Sociales y Ecónmicos." Quito.

Conaghan, Catherine M. (1984). "Redemocratization and the Party System in Ecuador." Paper presented at the Middle Atlantic Council of Latin American Studies, April 5-7, 1984.

CUNA (Consejo Unitario Nacional Agrario) (1985). Los Partidos y el Agro. Lima: CEPES.

Dandler, Jorge and Colin Sage (1985). "What Is Happening to Andean Potatoes? A View from the Grass-roots," Development Dialogue. NO. 1 (1985): 125-138.

de Janvry, Alain (1981). The Agrarian Question and Reformism in Latin America. Baltimore: The Johns Hopkins University Press.

Dourojeanni Ricardi, M., et al (n.d.). Proceso Agrario: Hacia Donde? Serie Intercampus de la Universidad del Pacíco, No. 13 (Probably late 1984 or early 1985).

Eguren, Fernando (1980). "Política Agraria vs. Produccion de Alimentos?" QueHacer, No. 3 (March): 34-41.

_____ (1982a). "El Mensaje del Sr. Ulloa y Las Lluvias," QueHacer, No. 15 (February): 22-27.

_____ (ed.) (1982b). Situación Actual y Perspectivas del Problema Agrario en el Perú. Lima: DESCO.

_____ and Alfredo Filomeno (1982). "Dos Congresos, ¿un Solo Camino?" QueHacer, No. 18 (August): 48-59.

ENCI (Empresa Nacional de Comercialización de Insumos) (1977). "Memoria Anual." Lima, mimeo.

FAO (Food and Agricultural Organization) (1982). Production Yearbook. Vol. 35. Rome: United Nations.

_____ (1985). Production Yearbook. Vol. 38. Rome: United Nations.

Fernandez, Gaston and Schodt, David (1984). "Participation and Reform in Liberal Democracies: A Comparative Study of Colombia and Ecuador." Paper presented to the Midwest Political Science Association, Chicago.

Fernández Jorge (1982). "La Producción de Alimentos en el Perú Un Problema Indegesto," QueHacer, No. 17 (June): 86-94.

Figallo, Flavio (1984). "Cuatro Tesis Equivocadas sobre Las Parcelaciones," QueHacer, No. 28 (April): 46-50.

Figueroa, Adolfo (1984). Capitalist Development and the Peasant Economy in Peru. New York: Cambridge University Press.

Figueroa Arevalo, Adolfo and Javier Portocarrero Maisch. eds. (1986). Priorización y Desarrollo del Sector Agrario en el Perú. Lima: Universidad Católica del Peru and Fundacion Friedrich Ebert.

FitzGerald, E.V.K. (1976). The State and Economic Development: Peru since 1968. New York: Cambridge University Press.

_____ (1979). The Political Economy of Peru 1956-1978. Cambridge: Cambridge University Press.

FLACSO (n.d.). Elecciones en Ecuador 1978-1980. Quito: Editorial Oveja Negra.

García, Bertha (1976). "Formas Actuales de Organizacion y Accion Politica del Campesinado Ecuatoriano," Revista Paraguaya de Sociologia, Vol. 13, No. 37 (September - December).

Gilbert, Dennis (1977). The Oligarchy and the Old Regime in Peru. Ithaca, New York: Cornell University Latin American Studies Program Dissertation Series.

Gonzales, Alberto and German Torre (1985). Las Parcelaciones de las Cooperativas Agrarias del Perú. Chiclayo: Centro de Estudios Sociales "Solidaridad".

Gonzales Zúñiga, Alberto G. (1985). "Cooperativismo Agrario y Parcelacíon en la Costa del Peru." In Gonzales and Torre (1985): 75-142.

Grindle, Merilee S. (1986). State and Countryside: Development Policy and Agrarian Politics in Latin America. Baltimore: Johns Hopkins University Press.

Guerra Tovar, Julio (1986). "Alternativas de Inversíon Publica." In Figueroa Arevalo and Portocarrero Maisch, eds. (1986): 145-162.

Handleman, Howard (1979). "Ecuador: A New Political Direction." American Universities Field Staff Reports No. 47 (South America).

____ (1984). "The Dilemma of Ecuadorian Democracy," Universities Field Staff International Reports, Nos. 34, 35 and 36.

____ (1985). "Elite Interest Groups under Military and Democratic Regimes: Ecuador, 1972-1984." Paper presented at the Latin American Studies Association meeting in Albuquerque, New Mexico (April).

Hopkins, Raymond F. (1986). "Food Security, Policy Options, and the Evolution of State Responsibility." In Tullis and Hollist (1986): 3-36.

IDB (Inter-American Development Bank) (1985). Economic and Social Progress in Latin America: 1985 Report. Washington, D.C.

____ (1986). Economic and Social Progress in Latin America: 1986 Report. Washington, D.C.

IEE and FEPP (Instituto de Estudios Ecuatorianos and Fondo Ecuatoriano Populorum Progressio) (1985). Políticas

Estatales y Organización Popular. Quito: Editorial Mendieta.

INCE (Instituto Nacional de Estadística y Censos) (1982). "Indice de Precios al Consumidor Area Urbana," No. 8 (November).

INP (National Planning Institute) (1985). Public Investment Data, 1970-1980. Lima: Dirección General de Inversiones.

Instituto Nacional de Planificación (1975). Plan Nacional de Desarrollo 1975-1978. Lima.

Jaquette, Jane (1971). The Politics of Development in Peru. Ph.D. Dissertation, Cornell University.

JUNAPLA (Junta Nacional de Planificación y Coordinación Económica) (1973). Plan Integral de Transformación y Desarrollo 1973-77. Editorial Santo Domingo: Quito.

____ (1977). Evaluaci Plan Integral de Transformaci Desarrollo 1973-1977. Quito.

____ (1978). Estadísticas Financieras de la Inversion Real Publica per Sectores 1970-1977. (October).

____ (n.d.). Evaluacíon: Políticas, Objetivos y Metas; Plan Integral de Transformación Desarrollo 1973-77. Quito.

Lajo Lazo, Manuel (1986). La Reforma Agroalimentaria: Antecedentes, Estrategia, y Contenido. Cuzco: Centro de Estudios Rurales Andinos.

Lee, Rensselaer W. III (1985-1986). "The Latin American Drug Connection," Foreign Policy, No. 61: 142-159.

Lefeber, Louis. ed. (1985). La Economía Política del Ecuador: Campo, Region, Nación. Quito: Corporación Editora Nacional.

Malloy, James M. (1982). "Peru's Troubled Return to Democratic Government." Universities Field Staff International Report, No. 15.

Martínez, Daniel (1985). "Política Económica Agraria del Nuevo Gobierno," Socialismo y Participación, No. 32 (December).

_____ and Armando Tealdo (1982). El Agro Peruano 1970-1980: Analisis y Perspectivas. Lima: CEDEP.

_____ (1985). "Desarrollo de la Producción Alimentaria: Una Estrategia," Socialismo y Participación, Vol. 30 (June): 1-22.

Martz, John D. (1983). "Populist Leadership and the Party Caudillo: Ecuador and the CFP, 1962-1981," Studies in Comparative International Development, Vol. XVIII, No. 3 (Fall).

_____ (forthcoming). The Elections in Ecuador. Washington, D.C.: American Enterprise Institute.

McClintock, Cynthia (1981). Peasant Cooperatives and Political Change in Peru. Princeton, N.J.: Princeton University Press.

_____ (1984). "Why Peasants Rebel: The Case of Peru's Sendero Luminoso." World Politics, Vol. XXXVII, No. 1 (October): 48-84.

Mendez, María Julia (1982). "Cooperativas Agrarias: La Libertad de Eligir lo que a Ti Me Conviene," QueHacer, No. 15 (February): 28-36.

Minaya, José (1986). "La Producción Agroindustrial en el Valle de Viru: El Caso del Esparrago." Unpublished paper, Department of Social Anthropology, University in Trujillo.

Ministerio de Economía y Finanzas (1986). El Presupuesto del Sector Público para 1986. Lima.

Ministry of Agriculture (1985). "Situación Actual del Agro." Lima, mimeograph.

Ministry of Agriculture and Livestock (Ecuador) (1983). "Inventario Nacional de Proyectos en Ejecución del Sector Agropecuario." Dirección Sectorial de Planificación. Quito. Mimeo.

NACLA (1975). "Ecuador: Oil Up for Grabs," NACLA's Latin America and Empire Report, Vol. IX, No. 8 (November).

OAS (1984). Statistical Bulletin, Vol. 6, No. 1-4 (January-December).

Palmer, David Scott (1984). "The Changing Political Economy of Peru under Military and Civilian Rule," Inter-American Economic Affairs, Vol. 37, No. 4 (Spring): 37-62.

Pas, Luis, et. al (1982). "Debate Agrario," Socialismo y Participación No. 18 (June): 19-38.

Pease, Henry (1977). El Ocaso del Poder Oligárquico. Lima: DESCO.

Portocarrero, Felipe M. (1982). "The Peruvian Public Investment Programme, 1968-1978," Journal of Latin American Studies, Vol. 14, No. 2: 433-45.

_____ (1983). Inversión Pública y Gestión Economica. Lima: La Fundación Friedrich Ebert.

Posner, J.L. and McPherson, M.F. (1982). "Agriculture on the Steep Slopes of Tropical America; Current Situation and Prospects for the Year 2000," World Development, Vol. 10, No. 5 (May): 341-353.

Presidency of the Republic (1981). Peru 1981. Lima; Presidency of the Republic.

_____ (1982). Peru 1982. Lima: Presidency of the Republic.

_____ (1984). Peru 1984. Lima: Presidency of the Republic.

Rojas Senisse, Hugo (1984). Azucar: Crisis y Alternativa. Lima: Instituto de Apoyo Agrario.

Rosero García, Fernando (1983). "Las Camaras de Agricultura y la Política Agraria del Ministro Vallejo." Unpublished paper, Social Sciences, Catholic University.

Ruff, Samuel (1984). "Agricultural Progress in Ecuador, 1970-1982." Foreign Agricultural Economic Report Number 208, United States Department of Agriculture.

Salcedo, José María (1984). "Paro Agrario: Apagones, Fogatas, y Arboles Caídos," QueHacer, No. 28 (April): 35-45.

Sanderson, Steven E. (1986). "The Emergence of the 'World' Steer: Internationalization and Foreign Domination in Latin American Cattle Production." In Tullis and Hollist (1986): 123-148.

Scott, Cristopher D. (1981). "Agrarian Reform and Seasonal Employment in Coastal Peruvian Agriculture," Journal of Development Studies, Vol. 17, No. 4 (July): 282-306.

Thorp, Rosemary and Geoffrey Bertram (1978). Peru 1890-1977: Growth and Policy in an Open Economy. New York: Columbia University Press.

Tullis, F. LaMond and W. Ladd Hollist (1986). Food, the State, and International Political Economy: Dilemmas of Developing Countries. Lincoln: University of Nebraska Press.

Universidad del Pacífico (1980). Peru 1980: Elecciones y Planes del Gobierno. Lima: Centro de Investigacíon de la Universidad del Pacifico.

Urban, Klaus (1986). "Irrigacíon y Desarrollo: Experiencias con Grandes Irrigaciones en la Costa Peruana." In Figueroa Arevalo and Portocarrero Maisch, eds. (1986): 163-201.

U.S. Presidential Agricultural Mission to Ecuador (1985). "Report of the U.S. Presidential Agricultural Mission to Ecuador." Agency for International Development (eds.).

U.S. Presidential Agricultural Task Force (1982). "Report of the United States Presidential Agricultural Task Force to Peru." AID (April).

Vidal CobíanAna Maria (1985). "La Legalizacíon de la Parcelacíon en las CAPs." In Gonzales and Torre (1985): 177-191.

Villalobos, Fabio (1985). "Ecuador: Industrialización Empleo y Distribución del Ingreso: 1970-1978." In Louis Lefeber. ed. La Economía Política del Ecuador. Quito: Corporación Editora Nacional: 243-293.

Werlich, David P. (1981). "Peru: Encore for Belaunde," Current History (February).

Wilson, Patricia A. and Carol Wise (1986). "The Regional Implications of Public Investment in Peru, 1968-1983," Latin American Research Review, Vol. XXI, No. 2: 93-116.

World Bank (1979). Ecuador: Development Problems and Prospects. Washington, D.C.

_____ (1981). Peru: Major Development Policy Issues and Recommendations. Washington, D.C.

_____ (1983). World Tables. Two Volumes. Washington, D.C.: World Bank.

5

President Marcos, Multinationals, the World Bank, and the U.S. Government: Domestic and International Political Economy of Philippines' Coconut Industry

Gary Hawes and Gretchen Casper

During the first few years after martial law was declared in the Philippines in 1972 the future looked bright. It was predicted that the country would be the next economic miracle of Asia. The economy was growing rapidly, attracting significant new foreign investment, and was breaking into the world market for low wage, labor-intensive manufactured goods such as textiles, clothing, and semiconductors.

Among the reasons for this optimism were the power of President Ferdinand Marcos and the new sense of discipline and direction martial law gave the country. Marcos had given authority to a group of technocrats, who would plan the nation's path to economic development, and to the military, which had been given free reign to impose order and to prevent strikes or economic disruptions.

The economic planners had begun implementing policies which would shift incentives away from producers for the domestic market and toward producers for the world market. In support of this policy shift, the World Bank and other lending or donor institutions were very generous with capital for the Philippines. This capital was, in turn, invested in infrastructure projects to support foreign investors, who needed export-processing zones, with the most up-to-date highways, ports, communications, waterworks, and electrical power grids. The capital also supported land reform and the green revolution, which together were designed to reduce agrarian unrest in the rice-growing regions, turn the country into a rice exporter, and in the process help keep labor costs among the lowest in Southeast Asia. It was also assumed that the agricultural sector would aid industrialization by creating from its profits a substantial amount of capital which would supplement international funds being invested into industry.

In short, the Philippines had all the characteristics of acountry on the verge of a breakthrough. It was a strong state with a strong leader which had the support of the international community. It was embarking on a process of export-oriented industrialization and following in the footsteps of its Asian neighbors---South Korea, Taiwan, Hong Kong, and Singapore.

Today under President Corazon Aquino, the Philippines is trying to recover from the economic crisis created by the Marcos regime. In the successful Pacific Basin region, the Philippines stands out as one of the few countries becoming economically worse rather than better off. The Philippines' gross national product actually shrank, decreasing in 1984 by 5 percent and in 1985 by 4 percent. Per capita income has dropped 14 percent since 1983; in the same time frame, exports also dropped 14 percent (Villegas, 1986). The Philippines, with the full support of the World Bank, has retreated to a model of economic development based on increasing agricultural exports (<u>Far Eastern Economic Review</u>, January 31, 1985).

If the earlier optimism about the Philippines was mistaken, it is easy to understand why. The 1970s was a period when a number of countries (e.g., Brazil, South Korea, Taiwan, Singapore, Mexico) were showing that it was possible through concerted effort and strong state action to break out of the colonial pattern of trade and into the ranks of exporters of manufactured goods. In fact, it was becoming almost axiomatic that finding a niche in the international economy required a strong state to shape the path of investment, create world-scale companies, and provide the initial subsidies which would allow a new industry to compete. Or, if the industrialization was to be based on foreign investment, a strong, stable state was necessary to provide the conditions which would attract the transnational corporations.

However, our study of the Philippines leads us to question whether the theory that strong states are key actors in the process of industrialization can be generalized to other countries. In the Philippines, development policy was implemented in such a way that not only did the economy shrink, but a political crisis was created leading to the downfall of the president. Is a strong state which is committed to reshaping the economy, stimulating the export of manufactured goods, and rapid economic growth sufficient to guarantee success?

Most case studies to date have dealt with countries which have had at least a limited degree of success. The close correlation between strong states and rapid economic growth, such as in South Korea or Brazil, could mistakenly lead us to assume a casual relationship. What of the cases which have failed? What was there about the domestic political economy, the international environment, or the interplay of the two which led to failure? Are there pitfalls along the path to export-led industrialization which may prevent some countries from successfully making the transition?

This paper analyzes how a country with such economic potential as the Philippines would see its economy deteriorate rather than expand. Paradoxical as it may seem, we feel that the best way to learn why the Philippines has failed to live up to its billing as the next economic miracle of Asia is to study one of the nation's traditional export industries. As we hope to make clear below, the coconut industry is a very good case study to show the role of the state in the Philippines, its entrepreneurial nature, its political and economic interests, and, ultimately, which class segments the state represented.

THE COCONUT INDUSTRY

The Philippines is the world's largest producer and supplier of coconut products. In 1976, the country supplied 82 percent of the world's total requirements for coconut products. To produce this much for the world market, 86 percent of all coconut production that year went to exports and only 14 percent was consumed locally (Republic of the Philippines, Philippine Coconut Authority, n.d.). By most standards, the coconut industry is, next to rice, the largest and most important industry for the country. In 1978, of the 11,749,300 hectares (one hectare equals 2.47 acres) planted with agricultural crops, 24.6 percent (or 2,889,800 hectares) were devoted to coconuts. Of the 3.425 billion dollars which the Philippines earned in exports, 26.5 percent (or 908 million dollars) came from the export of coconut products (Republic of the Philippines, National Economic and Development Authority, 1980).

Fully one-fourth and, perhaps, as much as one-third of the country's population depend for the bulk of their income, either directly or indirectly, on the coconut industry. One scholar estimates that in production alone there are 250,000 landowners and caretakers, 50,000 owner-farmers, 500,000 tenants, and 1,000,000 farmworkers. This total of two million producers, when multiplied by the average family size---six---gives a total of twelve million Filipinos directly dependent on coconut production. To this must be added the thousands of traders in the marketing chain, the workers in the oil mills and desiccating factories, and the management employees (David, 1977).

The coconut palm is an indigenous tropical plant that has probably always served an important part in the subsistence needs of the Filipinos. The export of copra, however, did not begin until late in the nineteenth century when the Western European and U.S. manufacturers of soaps and margarine began turning to tropical oils as a source of raw materials. Among the first coconut oil mills established in the Philippines were Proctor and Gamble Manufacturing Corporation, a fully-owned subsidiary of the Proctor and Gamble Company, USA, and Philippine Refining Company, which operated under the control of Lever Brothers Company, of Cambridge, Massachusetts, and formed part of the U.S. branch of the British firm, Unilever, (Private Development Corporation of the Philippines, 1978).

World War I stimulated a boom in the coconut industry, since the oil produced from crushing the meaty part of the coconut is high in glycerine and thus found a ready use in the production of explosives for the war. World War II and the occupation of the Philippines by the Japanese resulted in the closure or destruction of most of the nation's coconut oil mills.

The post-World War II era was a time of future growth and diversification of the industry. In 1970, the Philippines produced 66.3 percent of the world's coconut oil exports and this rose to 72.8 percent in 1975. There were about twelve million Filipinos involved in coconut farming. However, ownership of the mills which turned the raw materials into exportable oil was farmore limited. In the mid-1970s, there were only fifty-one mills operating. Several of the largest mills were owned by Americans and others were owned by Chinese (either nationals of Taiwan, the People's Republic of China, or Hong Kong or ethnic Chinese of Filipino citizenship who, although citizens, are considered "aliens" by Filipinos). The production of desiccated coconut for the world's bakeries is dominated by U.S. and British firms located in the Philippines.

Thus, the coconut industry is crucial to the Philippines for several reasons. First, it affects a significant percentage of the population. Second, it is one of the country's most successful export crops, in terms of attracting foreign investments and earning foreign exchange. Third, since most of the production stage for the industry was foreign-owned, it made sense, initially, to rationalize the industry by Filipinizing it. However, as we will see below, this policy was chosen haphazardly and was not implemented with the aim of profitting the industry, or the economy, as a whole.

THE TAKEOVER OF THE COCONUT INDUSTRY

Starting in late 1973, the world experienced a boom in the price of most commodities, among them coconut oil. In the Philippines, this commodity boom was a tremendous windfall for producers, processors, and the nation as a whole, since export receipts skyrocketed. (In fact, 1973 was the last year in which the Philippines experienced a positive balance of trade in its national accounts.) Unfortunately for Philippine consumers, coconut oil also serves as the nation's staple cooking oil.

On August 20, 1973, in response to the huge increases in the price of cooking oil on the domestic market, the president issued Presidential Decree 276. This decree gave a government agency, the Philippine Coconut Authority, the power to place a levy of fifteen pesos (then roughly two dollars) on every 100 kilos of copra at the level of initial sale. This was the start of the Coconut Consumers Stabilization Fund and was to be a temporary levy of only one year's duration. The funds raised were to be used to subsidize the domestic price of coconut oil. Any monies remaining in the fund after the subsidies were paid to the producers would be the property of the coconut farmers.

In 1973, the membership of the Philippine Coconut Authority was reorganized. Most of the representatives from government were removed from the board and replaced with close political associates of the president. Its primary purpose became to collect and oversee the coconut levy.

The government soon realized that the levy was a relatively easy way to raise huge amounts of funds. In fact, annual levy collections averaged one billion pesos (Far Eastern Economic Review, January 8, 1982). The president raised the amount of the levy and expanded the uses of the funds. The Coconut Industry Development Fund was created in 1974 to finance a replanting project involving the phasing out of the native coconut trees and their replacement with a hybrid variety. The replanting schemewas necessary because many of the coconut trees had passed their prime production levels. The Philippine Coconut Authority created a program in which it purchased the seedlings with funds from the Coconut Industry Development Fund and distributed the seedlings to the coconut farmers. The seedlings were bought exclusively from Agricultural Investment Incorporated, owned by Eduardo Cojuangco, Jr.

All in all, there were several projects other than the original subsidy for which the levy funds were used. Support programs including a scholarship for the college education of coconut farmers' children and a life insurance company were created. Funds were also contributed to the Federation of Coconut Farmers, toward its operating expenses.

One of the most important tasks which the levy took on was the rationalization of the industry "to strengthen the country's bargaining power in world markets by centralizing coconut-oil trading" (Far Eastern Economic Review, January 8, 1982). The idea was to create a vertical integration of the industry under Filipino control. In implementing this policy, the Philippines was hoping for an OPEC-type organization which would allow it to raise and control the price of coconut oil. As it was assumed that the profits from this rationalization of the coconut industry would be invested in the country's industrialization process, foreign actors such as the World Bank and the U.S. gave their tacit support.

One stage of the rationalization policy was the forced buyouts of local coconut operations of foreign firms by Filipino businessmen close to the President. Presidential Decree 1468, issued in 1978, provided the means for the takeover of the coconut mills by the Philippine Coconut Authority. The Philippine Coconut Authority needed a way to convince owners of profitable coconut mills to sell them, since it would be easy to purchase the less profitable ones. What Presidential Decree 1468 did was to stipulate that rather than paying subsidies to all coconut oil mills which were selling coconut oil to the public, only those oil mills owned or controlled by the coconut farmers could receive subsidies from the Coconut Consumers Stabilization Fund. The funds from the levy were diverted to a new fund, known as the Coconut Industry Investment Fund. After Presidential Decree 1468 was issued, the subsidy to oil mills could go only to those mills which had been purchased by the bank which the Philippine Coconut Authority had acquired as the central bank for the coconut industry, since the understanding was that the coconut farmers owned the bank.

Oil mills, refining plants, and export firms were purchased with levy funds, with the understanding that they would be held for the farmers. United Coconut Oil Mills was formed and quickly controlled 93 percent of coconut milling operations in the country. This corporation was also given a virtual monopoly on exporting coconut products and came to control 80 percent of coconut exports (Far Eastern Economic Review, January 8, 1982). Because United Coconut Oil Mills was able to control both buying the copra at the local level and selling the oil abroad, it was able to deflate the domestic market prices of copra by 9 to 15 percent (de Dios, 1984).

The coconut industry's bank, the United Coconut Planters Bank, had also been purchased from levy funds. All levy funds were deposited at United Coconut Planters Bank interest-free. Shares in the bank were to be distributed among the coconut farmers, since 72 percent of the bank was purchased with levy funds. However, actual distribution of farmers' shares was slow, and the nondistributed shares were controlled by Eduardo Cojuangco, Jr. "in the name of" the farmers.

In short, what the president had done was to completely takeover the coconut industry. First, a levy was placed on the sale of coconut products to subsidize the high cost of cooking oil. In the years after 1979, the levy proceeds were used to, in the words of the President, "rationalize" the industry. This involved the creation of a set of interlocking directorates through which the president's associates controlled the Philippine Coconut Authority, the United Coconut Oil Mills, the United Coconut Planters Bank, and the Agricultural Investment Incorporated. Most prominent among these presidential associates were Cojuangco, regional director of the ruling Kilusang Bagong Lipunan (New Society Movement) party for politically-sensitive Central Luzon, and Juan Ponce Enrile, Minister of National Defense.

Once the institutional machinery, funding, and legal groundwork were all in place, the new powers in the industry were able to squeeze out foreign investors and local Filipino capitalists from the coconut oil milling industry. For example, both Granexport Manufacturing Corporation, a subsidy of the U.S.-based Cargill Corporation, and Legaspi Oil Mills, the joint-venture subsidiary of Japan's Mitsubishi and the Philippines' Ayala Corporation, were forced to sell the irinvestments in the coconut oil milling industry to the quasi-public United Coconut Oil Mills.[1] Representatives of the foreign investors were unanimous in saying that prior to Presidential Decree 1468, they had no intention of withdrawing their investment from the industry or the country. They also all agreed, however, that they had received a fair price when they sold out.

Politically, the takeover was equally important. An industry which has an impact on at least one-third of the Filipinos was directly under the control of the President's men and no longer in the hands of potential opponents---coconut landlords, foreign investors, or Chinese businessmen. Also, a tremendous source of patronage and political funds was made available. By 1980, the levy had generated a total of 5.95 billion pesos (slightly more than 700 million dollars) (Business Day, September 1, 1980).

These funds were used in myriad ways, some of which were directly linked to political legitimation. There were, for example, the President Marcos and Defense Minister Enrile scholarships for the deserving children of coconut farmers, there were also life insurance and replanting schemes---all widely advertised, all to benefit the small farmers, it was alleged, and all far more important as political rhetoric than in actual implementation. The fact is, the industry was taken over in the name of coconut producers using funds levied from the producers, but the vast majority of the proceeds went to finance the political and economic goals of President Marcos and his associates.[2]

THE IMPACT OF THE TAKEOVER

By 1980, the international market for coconut oil had changed dramatically. In response to the invasion of Afghanistan, the U.S. placed an embargo on agricultural shipments to the Soviet Union. This action reduced demand in the U.S. for vegetable oil at a time when soybean and coconut supply levels were high. Coconut prices dropped 33 percent from 1979 to 1980 (Far Eastern Economic Review, January 8, 1982). This drop in the market price for coconut products greatly affected coconut farmers and tenants. Although extensive data are not available to analyze indepth the effect of the coconut levy on the farmers, there are available figures regarding the economic misery of the coconut farmers. By the early 1980s, the coconut levy was consuming 25 percent of coconut farmers' income. Their net incomes ranged between 5,500-5,900 pesos, while the subsistence level was 9,600 pesos (Far Eastern Economic Review, January 8, 1982). In comparison, the United Coconut Oil Mills had a large source of capital---the coconut levy---that was protected from market fluctuations.

When the coconut industry takeover was initiated in the mid-1970s, the reaction of Filipinos and foreign investors was muted. Businessmen accepted the idea of rationalizing the industry. Also, while the market price had been deflated by the monopoly, the farmers were still receiving a higher price than in previous years. However, protest erupted when the world market price for copra dropped but the farmers still had to pay the levy.

The drop in price for coconut oil sparked a national debate on the coconut levy, especially the way in which the Philippine Coconut Authority had transferred a large percentage of the proceeds from the levy to a privately owned bank for the purchase of privately owned oil mills, which were then grouped under the umbrella corporation United Coconut Oil Mills. Once the funds from the levy were transferred from the government agency to the private agency, they were no longer subject to public scrutiny by the government's Commission on Audit. To this day, no clear accounting of the use of these publicly generated funds has been made. The secrecy surrounding all these maneuvers generated tremendous public cynicism and suspicion.

In 1982, the debate came out in the open in the Interim Batasang Pambansa (Interim National Assembly). The major proponent for the levy was Defense Minister Juan Ponce Enrile. Supporters of the levy stressed various programs which the funds bankrolled, including the scholarships and life insurance, as well as the farmers' ownership of the United Coconut Planters Bank.

One of the major opponents of the levy was Assemblyman Emmanuel Pelaez (an ex-senator and ex-vice president), who argued that the levy taxed coconut farmers more than what they actually received in benefits. The scholarship fund was only a symbolic program, offering a minimal number of grants annually. It was argued that the replanting program was potentially dangerous to the coconut industry. There was serious debate among agricultural specialists over the sturdiness of the hybrid, especially in resisting monsoons. One of the most serious charges was that the different support programs financed by the levy were run by interlocking directorates, thus setting up a monopoly in the coconut industry, which used farmers' money but benefited only the directors of the various corporations. Not surprisingly, technocrats, such as Prime Minister Cesar Virata and economists at the Development Academy of the Philippines, suggested ending the levy.

The open debate was exacerbated by President Marcos' flip-flopping on the issue. To stem the controversy, he suspended the levy for a six-month period during which studies were to be carried out to determine what action would be taken. During this period, United Coconut Oil Mills mounted tremendous political pressure by refusing to buy copra, arguing that there was an oil glut. The crisis worsened; coconut farmers, via a United Coconut Oil Mills' lobby effort, pressed the president for reinstatement of the levy. While Virata was out of the country at an International Monetary Fund conference, the levy was reinstated.

The crisis created by the decrease in the world market price of coconuts and the subsequent suspicions regarding the levy funds fueled political unrest in the country. Demonstrations and anti-government newspaper articles increased, all attacking the government's support of "crony capitalism."[3] There was speculation that these monopolies were only managed, and not owned, by these friends---that, in fact, they were only frontmen for the First Family.

There was also an increase in guerrilla activity in major coconut-producing areas. The New People's Army and the Moro National Liberation Front were taking advantage of the situation by informing tenants that there was a 1.70 peso difference (about twenty cents) for every kilo of coconut products sold between the price paid the farmer and the world market price, and that this surplus was going to President Marcos, with Enrile and Cojuangco being only the "collection agents."

By the early 1980s, the Marcos regime had been in power for fifteen years. It was an increasingly unpopular regime, beset by growing opposition in the cities and in the rural hinterlands. The coconut industry had been used to solidify the regime throughout most of the 1970s, rather than to fund investment projects in the economy. Now in the 1980s with declining prices for Philippine exports, the continuing extraction of funds for political legitimation from the producers of coconut products took a larger share of their incomes. Marcos, no longer so firmly in political control of the country, was now open to international pressure.

GROWING INTERNATIONAL PRESSURE ON THE MARCOS REGIME

The economic situation in the Philippines declined steadily in the 1980s. There were several reasons, both economic and political, behind this decline. First, the global recession depressed commodity prices, thereby decreasing the Philippines' foreign exchange earnings, since it is a primary goods exporter. Second, inflation depressed domestic demand, thereby reducing domestic business activity. Third, the Philippines' revenue collection was declining, forcing it to turn to external borrowing, with its inherent high interest rates. This last act strained Philippine debt servicing, as its deficits were increasing. By 1983, foreign banks were wary of lending more money to the Philippines because of its debt burden and pessimistic economic future. To acquire needed funds, the Philippines was forced to use short-term loans, which added to its debt service burden.

There were also political reasons for the Philippines' economic decline. First, the scale of corruption in the Philippine government had become notorious, even compared withother "lax" countries. One estimate was that as much as 30 percent of public spending was being siphoned off for private pockets, "compared with about 10 percent in 'normal' developing countries" (Peagam, 1984). Added to the loss of revenue was the fact that most of this money was invested or deposited abroad. Second, an increasing number of Filipino businessmen were openly criticizing "crony capitalism." There was little incentive for businessmen to sacrifice in the interest of economic recovery if it was perceived that the benefits would go primarily to the cronies. There was also increasing suspicion among businessmen that the monopolies controlled by the cronies, including the coconut industry, were being used to raise large sums of money to be diverted to a political slush fund (Peagam, 1984).

By the middle of 1983, it became clear that the government would not be able to meet the target set under an International Monetary Fund standby agreement of reducing the overall deficit in the balance of payments from 1.2 billion dollars to 600 million dollars. Instead, the deficit had increased to 1.3 billion dollars. Uncertainty about the economic future of the country was reflected in the termination by foreign banks of about 700 million dollars in credit and the increase in capital flight out of the country. An International Monetary Fund standby credit facility also lapsed when the Philippines could not reach its targeted deficit level (Peagam, 1984).

In August 1983, President Marcos withdrew from public, ostensibly to begin working on a new book. Rumors circulated that he was undergoing serious treatment for lupus, a kidney disease. Political and economic uncertainty grew. When Senator Benigno Aquino returned to the Philippines, despite indirect warnings from the First Lady Imelda Marcos not to, he was assassinated at the airport.

An economic and political backlash was unleashed. More credit was cancelled by foreign banks, foreign investment in the Philippines dropped, tourism came to a halt, and capital flight dramatically increased. The Philippines, to show goodwill to the International Monetary Fund in order to receive an emergency loan, further devalued its currency from 11 to 14 pesos per dollar. However, the debt crisis escalated. The balance of trade deficits rose to two billion dollars and the Central Bank's dollar holdings dropped below 400 million dollars. By mid-October 1983, the Philippines declared a 90-day debt repayment moratorium and asked the International Monetary Fund to help negotiate debt rescheduling.

The assassination also triggered political unrest. For the first time, political demonstrations were held in Makati, Manila's business district. Political groups, such as GABRIELA and MABINI, formed among the professional and upper classes and held weekly rallies, including visits to the Interim Batasang Pambansa for redress. Political action was also sparked among other economic groups, as labor unions held their own rallies. The University of the Philippines held anti-Marcos demonstrations. Pro-opposition publications such as <u>Veritas</u> and the newsprint edition of <u>Mr. and Ms.</u> sprang up. The New People's Army was able to make inroads into 80 percent of the provinces. Their increased strength was seen as the result of the depressed economy. The New People's Army also took advantage of longstanding political issues---human rights violations, government corruption, inefficiency of government services, andcrony capitalism (U.S. Senate, Committee on Foreign Relations, 1984).

The May 1984 elections, unlike other elections held undermartial law, became "an opportunity to express popular dissatisfaction with the Marcos government" (U.S. Senate, Committee on Foreign Relations, 1984). Surprisingly, the opposition received 59 out of 183 elected seats in the regular Batasang Pambansa[4] with speculation that they would have won over 100 seats had the elections been fair. "In Metro Manila where Mrs. Marcos was in charge of the [New Society Movement's] campaign, the loss of 15 out of 21 seats to the opposition was a humiliation" (U.S. Senate, Committee on Foreign Relations, 1984).

With such widespread unrest, Marcos was forced to accept certain international demands in return for funds. The U.S. had been pressuring him to "break up his cronies' monopolies in the sugar and coconut-oil trade, to curb the dynastic ambitions of his wife, Imelda, and to prepare the wayfor an orderly transfer of power to future leaders" (<u>Asian Wall Street Journal</u>, January 9, 1984). The U.S. pressure on these items was in part a result of talks with Philippine businessmen. They were seen as necessary reforms before any serious regeneration of the Philippine economy could occur. The U.S. had targeted crony capitalism specifically to downgrade favoritism and instill businessmen's confidence in the economy.

The World Bank likewise attacked the government's management of the coconut industry. In the loan agreement between the Philippines and the World Bank, the latter specified that it would review the adequacy of progress in reforms of the Philippine Coconut Authority (<u>Far Eastern Economic Review</u>, January 31, 1985). The World Bank also suggested are organization of the Ministry of Agriculture and of the PCA, as well as a change in the government's coconut industry policies, especially the nature of the relationship between the government's regulatory agency and the processing industry and the rationale for the composition of the governing board of the regulatory agency (<u>Far Eastern Economic Review</u>, January 31,1985).

In January 1985, a presidential decree was issued allowing all coconut millers the right to export their product independently. To this end, the United Coconut Oil Mills was dismantled. However, the monopoly continued under Cojuangco, albeit in a different form. The coconut mills, which were originally owned by the United Coconut Oil Mills, entered into a cooperative. Thus, the domestic purchase price for copra continued to be depressed and controlled by Cojuangco. Because of this result and because of the general resistance of Marcos to significantly demonopolize agricultural sectors of the economy, the International Monetary Fund refused to release loan funds which were scheduled to be given to the Philippines in October 1985 (Villegas, 1986).

In response to mounting international pressure to prove his domestic support and, therefore, underscore his legitimacy to remain as president, Marcos declared a snap election, held in February 1986. He was counting on traditional electoral politics in the Philippines to reassert his rule. First, he was assuming a fractured and, therefore, weakened opposition. Second, he assumed that between vote buying, intimidation, and election fraud he would be able to construct an electoral victory. In an unusual turn, the opposition united behind Corazon Aquino, the widow of Benigno Aquino, Jr. The National Movement for Free Elections, an independent poll watcher, protected the ballot boxes and publicized to the foreign press in the Philippines the election fraud events. Even the Commission on Election's computer vote tallyers walked out to boycott the blatant vote tampering by the government.

Marcos claimed that he won the election and was inaugurated into the presidency. However, fearing arrest, Defense Minister Enrile and General Fidel Ramos proclaimed Aquino the legitimate winner of the election and defected from the Marcos government. In a quick succession of events, Marcos fled the country and Aquino was sworn into the presidency, promising economic and political reforms. Fourteen years after the declaration of martial law, the promises for both a strong state and a strong economy had been proven false.

REASSESSING THE ROLE OF THE STATE

Recent studies of the Third World state have focused on the role the state plays in the economy and have often used the nature of the state's intervention in the economy as a means of predicting success in industrialization. Generalizing from the success of a few newly industrializing countries, it has been argued that the state may accelerate the process of growth by choosing industries in which the nation will concentrate and channeling investment into those industries, by investing in the infrastructure which will make the nation an attractive place for investment by the private sector---whether that private sector be national or transnational---or by disciplining the populace, reducing political participation, and depressing the income levels of the work force to foster low-wage, labor-intensive industries (Evans, 1979; Cumings, 1984). Others, more interested in the class basis of these states than in the state's entrepreneurial role, have argued that the state intervenes in the economy because it has its own state interests to protect or, alternatively, because the state defends a particular set of class interests (O'Donnell, 1978; Cardoso and Faletto, 1979).

We want to argue that these two theoretical insights should be combined. In the case of the Philippines, a new coalition of class interests reshaped the nature of the Philippine state following the declaration of martial law. It was an interventionist state determined to industrialize based on low-wage, labor-intensive manufactures. We know something about the class segments supporting the state because we know which actors supported this new process of export-oriented industrialization. The World Bank, the International Monetary Fund, and the U.S. government were all firmly behind this new model of development, largely because the model is dependent on investment by transnational corporations and the free flow of capital across national borders. Domestically, the new model was supported by a segment of the local business class which had the resources and connections to enter into joint venture partnerships with the foreign investors. Similarly, a growing middle class of managers, lawyers, and accountants all benefited from the growing manufacturing sector. Thus, the state in the Philippines was clearly not just a neutral set of institutions processing the demands of political interest groups. Also, the state was definitely a class state since it rested so clearly on repression of the interests of workers.

As was in the case in other parts of the Third World, the Philippine state undertook an aggressive, interventionist role inthe economy. Whenever possible, the state's machinery was used to foster a process of export-oriented industrialization. In part, this development was funded by loans from the international commercial banks as well as from the World Bank. To cover the rest of the cost, the Philippines relied on the surplus generated by agricultural exports. One of the agricultural sectors used to support development was the coconut industry.

A virtual monopoly was formed within the coconut industry, with interlocking directorates for the different public and private corporations involved. Members of the Philippine Coconut Authority appointed by President Marcos controlled both the United Coconut Planters Bank, where the Authority-collected levy was deposited, and United Coconut Oil Mills, the umbrella corporation which had a monopoly on the refining and export of coconut products. These government actions centralized control of all the surplus generated by the coconut industry.

If we visualize the state as a coalition of class interests, then centralizing the control of surplus generated by the coconut industry could also be seen as a step towards redirecting the surplus out of agriculture and into a new export-oriented manufacturing sector which was the preferred model of development. This, we argue, was why the World Bank and the U.S. government were largely silent when the Philippine state used its power to squeeze foreign investors out of the coconut milling industry. It was expected that the coconut industry would be "rationalized" and that the surplus generated by the industry would be increased, centralized, and utilized to advance low-wage, labor-intensive manufacturing. If the state was to be the primary investor in the infrastructure necessary to attract foreign investment, then the state needed sources of funding. Taxation, the normal source of government funding, was a notoriously inefficient source of funds in the Philippines.

If we take President Marcos to be the leader of a political regime and that regimes can be differentiated from states, then it becomes apparent that his goals for the coconut industry were different from those of the more abstract political coalition we have called the state. Likewise, the coalition of class interests which supports a strong state role and a new model of development may be different from the political coalition which supported the continuation in office of Ferdinand Marcos. Marcos could enhance his own value by efficiently advancing the interests of the state. However, since the Philippine state is a class state, advancing the interests of the state would not always make him politically popular with the Filipino people. He, thus, had a personal, political interest in defending and legitimizing his regime. In short, there were cases where his personal political interests conflicted with the interests of the state.

We believe the coconut industry is a good example of where the political interests of the Marcos regime conflicted with the economic interests of the Philippine state. State power was used to take over the coconut industry and this included action detrimental to the interests of one segment of the bourgeoisie. While the evidence is not clear on why other elements of the bourgeoisie went along with this case of state intervention in the private sector, there are at least two possible explanations. First, the industry suffered from over-capacity in the oil milling sector, with a number of new mills sitting idle because of shortages of raw materials. Secondly, the production of coconuts had been suffering from declining yields because of aging stands of coconut palms and the very low level of capital intensity of production techniques. In short, the industry was in dire need of precisely what Marcos promised---rationalization.

However, Marcos used the surplus generated by the coconut industry largely for the legitimization of his regime rather than for reinvestment in the economy. This legitimization was successful for several years following the institution of the levy and the takeover of the privately owned mills. The public benefited from the subsidies on cooking oil and no one had reason to suspect that the surplus generated was not going for the revitalization of the coconut industry.

But, as our case study shows, by 1980 prices for coconut products were dropping rapidly. Suspicion and resentment created by the levy and monopoly began to spark an upsurge in domestic protest against the regime. In trying to defend itself, the regime's need for funds increased. As profits from the industry were channelled more and more into political projects and less and less was reinvested in infrastructure of the coconut industry itself, the economy stagnated. The obvious favoritism of the regime's actions alienated domestic business interests.

Domestic political opposition then opened the door for a reassertion of state interests, thereby further weakening the Marcos regime. The reassertion of state interests is exemplified by the pressure being exerted by the domestic business class, the World Bank, the International Monetary Fund, and the U.S. government for reforms which would reduce the personal power of Marcos, but at the same time would reemphasize export-oriented industrialization as a path to development.

We believe that the body of literature in which the Third World state is the unit of analysis provides a useful perspective for investigating the political economy of a group of countries often said to be newly industrializing. What has been particularly useful are the theoretical breakthroughs which differentiate between the state and the regime and which assign a relative degree of autonomy to the state. In the case of the Philippines, the state has demonstrated that autonomy---it was able to squeeze out of the coconut industry Mitsubishi, one of Japan's largest grain traders. This achievement was not the work of a state which is the instrument of the capitalist class or even the work of a state which simply responds to the demands of powerful economic interest groups.

The failure of the Philippines to successfully manage the transition to export-oriented industrialization indicates an important area in which the theory of the Third World state maybe refined. To focus exclusively on the role of the state and what the state may do in the way of intervention in the economy to foster industrialization is to overlook the importance of the regime. In many cases in the Third World, the costs of industrialization will make any regime politically unpopular. The existence of an unpopular regime makes it difficult to efficiently administer the process of industrialization. In this case study, it was, ultimately, the weakness of the Marcos regime which undermined the industrialization process.

Thus, we would conclude that looking at the strong state and its interaction with the international environment helps us to understand which countries are likely to begin industrializing. However, the analysis needs to be extended one step further to look more closely at the nature of the regime and its interaction with domestic political forces if we are to understand which countries are likely to successfully make the transition to export-oriented industrialization.

NOTES

1. In an interview with Enrique Zobel, president of Ayala Corporation, he pointed out that he had a "gut feeling" that the industry was going to be nationalized anyway and so he wanted to be the first one to sell out and thereby get the best deal. He also thought that those who were still in the industry were getting "screwed," that they would be bled to death, and that United Coconut Oil Mills controlled the supply of copra and their mills were exempt from paying the levy. He made it clear that it would be impossible to make a profit in that environment and that this was part of the United Coconut Oil Mill's design to force the private sector to sell their coconut oil mills.

2. An internal memorandum of the Development Academy of the Philippines dated 15 September 1981, reported that for the period 1973-1978, only 30 percent of all levy funds collected went to subsidize the cost of cooking oil.

3. Crony capitalism here refers to President Marcos' propensity for allowing his close friends to construct monopolies in different sector of the economy. The coconut monopoly was repeated in sugar, wheat imports, cigarette filters, and other areas where presidential friends received special favors.

4. In 1984, the National Assembly was convened as a regular body and, therefore, was no longer an interm legislature.

REFERENCES

Business Day (September 1, 1980).

Cardoso, Fernando Henrique and Enzo Faletto (1979). Dependency and Development in Latin America. Berkeley: University of California Press.

Cuming, Bruce (1984). "The Origins and Development of the Northeast Asian Political Economy." International Organization 38 (Winter): 1-40.

David, Virgilio (1977). The Barriers in the Development of the Philippine Coconut Industry. Unpublished MBA thesis, Ateneo de Manila University.

de Dios, Emmanuel S., ed. (1984). An Analysis of the Philippine Economic Crisis: A Workshop Report. Quezon City, Philippines: University of the Philippines Press.

Evans, Peter (1979). Dependent Development. Princeton: Princeton University Press.

Galang, Jose (1985). "Economic Husbandry: Manila Looks to Recovery Led by a Restructured Farm Sector." Far Eastern Economic Review (January 31): 46-49.

Lachica, Eduardo (1984). "US Presses Marcos to Curb Cronies, Prepare Succession." Asian Wall Street Journal (January 9).

O'Donnell, Guillermo A. (1978). "State and Alliances in Argentina, 1956-1976." The Journal of Development Studies 15 (October): 3-33.

Peagam, Norman (1984). "The Spectre that Haunts Marcos." Euromoney (April): 46-63.

Private Development Corporation of the Philippines (1978). The Philippine Coconut Oil Milling Industry. Manila: Private Development Corporation of the Philippines.

Republic of the Philippines, National Economic and Development Authority (1980). 1980 Philippines Statistical Yearbook. Manila: National Economic and Development Authority.

_____ (n.d.). Philippines Coconut Authority. <u>The Philippine Coconut Industry</u>. Quezo City, Philippines: Philippine Coconut Authority.

Sacerdoti, Guy (1982). "Cracks in the Coconut Shell." <u>Far Eastern Economic Review</u> (January 8): 42-45.

U.S. Senate, Committee on Foreign Relations (1984). "The Situation in the Philippines: Staff Report." Manuscript (September).

Villegas, Bernardo M. (1986). "The Philippines in 1985." <u>Asian Survey</u> XXVI (February): 127-140.

6

Problems of Agricultural Development in Turkey

Mahir Fisunoğlu and Birol A. Yeşilada

Agriculture has always been a crucial section of the Turkish economy. Even though this sector's share of the total GDP reached 18.0 percent in 1985, it accounted for 29.6 percent of Turkey's export earnings and provided employment for 59.1 percent of the country's labor force (SPO, 1986: 3, 34 and 77). Moreover, the comparison of these figures with the averages of other middle income less developed countries brings out the relative significance of agriculture for the Turkish economy. In 1984, this sector in other middle income states accounted for 14.0 percent of GDP, 23.0 percent of export earnings, and 44.0 percent of the total labor force (World Bank, 1986: appendices). Yet, despite the significance of agriculture for Turkey, this sector of the economy suffered from lack of land and agricultural reform. In essence, all official efforts aimed at promoting agrarian reform in Turkey have been slow and plagued with domestic and external constraints.

This chapter examines the problems of agricultural development in Turkey between 1963 and 1983. This period is important for three major reasons. First, following the 1960 military coup and the birth of the Second Republic in 1961, state planning in the form of Five Year Development Plans controlled economic development in Turkey. Since then, the governments have drafted five such macroeconomic plans. Second, prevailing development models divide this period into two subperiods. Between 1963 and 1980, the import-substitution model guided economic development, whereas since 24 January 1980 an export-led model has affected growth. And third, with the adoption of the January 1980 IMF sponsored economic austerity measures, the Turkish authorities have undertaken serious efforts to integrate Turkey's economy with the world capitalist market.

Thus, we will examine the impact of these growth models on the agricultural sector in Turkey and identify key factors that inhibit agricultural development.

CHANGES IN DEVELOPMENT POLICIES

As stated above, development policies during the Second Republic are presented in the form of Five Year Development Plans starting in 1963. An examination of the policies of various government coalitions during 1963-1983 indicates that the center-rightist Justice Party (JP) dominated policy making in these years. This, in turn, makes the JP largely responsible for the policy priorities of the civilian govnments during this time period.

The First Five Year Development Plan (1963-1967) gave the guidelines for economic development which the individual administrations were to implement through their annual programs. The First Plan voiced deep concern for economic growth and problems of social equity. It called for seven percent real growth in GNP and an increase in domestic investment to 18.3 percent of the GNP. It also suggested the reliance on external sources to finance part (3.5 percent) of these investments through credits and foreign aid in the form of grants. The domestic public and private sectors would provide the remaining portions. As for industrial growth, the main policy remained the import-substitution growth model. In the agricultural sector planners emphasized the use of chemicals, fertilizer, improved seeds, and mechanization to increase productivity. To overcome problems in social equity, such as unequal distribution of national income among social classes and regions of the country, the First Plan called for investment in the underdeveloped provinces but failed to emphasize any concrete reform policies. Officials assumed increased investment would improve income distribution in the country. Furthermore, to make investments in these regions attractive to the private sector, the government passed laws promoting tax incentives to the investors (Tokgöz, 1976). In essence, this first plan was not conceptually different from previous economic attempts during the greater part of the First Republic. It proposed no major structural economic changes. This was especially true for the agricultural sector.

The Second Five Year Development Plan (1968-1972), adopted during the JP rule, stressed economic growth and placed little emphasis on social justice and equity. It sought a high rate of economic growth, at seven percent annually, through further import-substitution measures to protect domestic industries. Again the Second Five Year Development Plan downgraded the role of agriculture. It called for increased production but failed to emphasize much needed agricultural reform. Concerning the problems in social equity, the Plan revised the concepts of the First Plan. The Second Plan (SPO, 1969: 3) did not emphasize regional development nor did it address the importance of income re-distribution in a concrete manner. It simply suggested equal opportunities for everyone. This policy strengthened the private sector through long term low-interest state loans to encourage private sector investments. To achieve this, the JP coalition in 1968 established the Teşvik Uygulama Dairesi (The Office of Encourgement for Investments) within the State Planning Organization to assess private sector investment proposals and to approve the necessary state credits for their implementation. It is crucial to note that almost all of these investment credits were targeted for the industrial sector of the economy.

During this period, however, the country experienced serious social unrest in major urban centers as student and labor groups began to question the JP government's policies on economic growth, social equity, and political rights of citizens. Soon the student and labor unrest turned into ideological confrontations, which were at times intensified by the administration's support of the rightist organizations (Kongar, 1981; Yeşilada, 1984). As the violence between the rightists and leftists worsened, the military high command issued a memorandum to the President of the Republic that forced the resignation of the JP government on March 12, 1971. For the next two years transition government governed Turkey under the direct observation of the Joint Chiefs.

The 1971 military coalition seemed at the beginning to favor a centrist administrative solution to Turkey's crisis. Professor Nihat Erim, a prominent member of the Republican Peoples Part (RPP), was asked to form the government. Erim established a reform-oriented cabinet and pledged to direct his efforts toward the restoration of law and order and the completion of long forgotten Kemalist reforms, including the land and agricultural reform. However, the center right coalition in the Assembly successfully brought about this government's resignation within a year by persuading the generals that the proposed reforms were unnecessary and communist oriented. The country was then run by conservative interim governments, including a second administration formed by Erim, under the patronage of the military until the 1973 national elections.

The military based coalitions adopted the Third Five Year Development Plan (1973-1977) which by and large reflected the views of the JP coalition in the Second Plan. While this Plan (SPO, 1973: 49-50) emphasized the need for economic growth, the plan also warned of any measures designed to change income distribution in the country. Government officials believed such measures would have a negative impact on the rate of economic growth. Similarly, this Plan stressed that any attempts taken to eradicate regional differences in socio-economic development may, in the long run, slow capital accumulation and growth by causing an ineffective use of resources. This meant the Plan viewed reform unnecessary for land distribution or the mode of agricultural production. It only suggested large landowners use Green Revolution technological advances. In essence, the Plan argued that priorities in the choice of location for economic investments, especially by the private sector, should be based on the economic criteria of minimizing costs.

Following the 1973 national elections and return to civilian rule, the various development policies of coalition governments between 1973 and 1980 reflected compromises among the coalition partners for the political gains of each party. An examination of the differing policy stands of these parties demonstrates the complexity of various coalitions (four in all), formed during this period.

The social democrats of the RPP favored economic development policies aimed at both ending the existing inequality between urban and rural areas as well as creating the needed capital to support growing industry within the import-substitution model (Cumhuriyet Halk Partisi, 1976: 26). The RPP coalition believed that growth in Turkey's largely agrarian economy depended on rural cooperative enterprises composed of small and medium size farms engaged in animal husbandry, poultry, fisheries, and mechanized farming. Through this plan, these industries would compete in internal and external markets providing surplus capital for the economy. The utilization of surplus capital would then further strengthen the industrial sector of the economy.

The conservative laissez-faire coalition of the Justice Party, however, continued to argue in favor of classical economic liberalism. The coalition emphasized an open economy for foreign capital as the necessary tool to develop the country and to provide jobs. It further supported the private sector and the limitation of state investments to only areas of activities deemed unprofitable or too expensive for the private entrepreneurs. The party program in 1973 and 1977 included no new ideas from those specified in the Second Plan (Adalet Partisi, 1973, 1977).

Thus, the two major rivals of Turkish policies held vastly differing views of the economic priorities of the country. The RPP elite emphasized the need for reform in the agricultural sector that would profit both small and medium-size farms and the economy as a whole. The social democrats realized that reliance only on industrialization did not suffice to bring the Turkish economy to par with those of the EEC countries. The agricultural sector also needed serious reforms. On the other hand, the conservative JP elite failed to see this need. The conservatives emphasized industrial growth by favoring entrepreneurs and downplayed the role of agriculture in this process. For them, the traditional feudal system sufficiently generated agricultural surplus for exports.

The Fourth Five Year Development Plan (1979-1983) was adopted during the rule of the rightist Nationalist Coalition. This Plan vividly highlights this group's ineffectiveness in dealing with the country's crucial problems. The only real difference between this plan and the previous three was the attempt to analyze social equity concerns in more detail. A long discussion of income distribution in Turkey was provided but no concrete policy suggestions were given. Furthermore, there was no discussion of land inequality in the country (SPO, 1979: 433-478). Simply, the presence of four parties in the nationalist coalition, with distinct views on development, hindered any real compromise except in abstract and vague terms. During their stay in office, the nationalist coalitions continuously relied on heavy external borrowing to meet trade deficits while failing to adopt policies to handle the growing economic and political crises. The reasons behind the economic crises were twofold.

Following the 1973 oil crisis, Turkey's oil import expenditures jumped drastically; while the total oil bill in 1973 was about 17 percent of export earnings, in 1978 it had reached 76 percent, or a total of $1.4 billion. Starting in 1974, short-term commercial credits financed, to a large extent, the resulting trade deficits. However, by 1979 Turkey increasingly found these credits difficult to service. Nevertheless, the Nationalist Coalition's unwillingness to alter development policies in the face of increasing economic troubles remained the major problem. It maintained investment demands at high levels to meet plan strategies and ignored danger signals in the sectoral performance. Also, attempts were not made to increase domestic savings. Likewise, the balance of payments policies remained inward looking, favoring import-substitution over export orientation. By 1975, Turkey had clearly satisfied the earlier stages of the import-substitution growth model---the completion of the domestic consumer market oriented industrialization. The next stage was the expansion of industrial growth through export-oriented policies. Such policies would then help reduce the trade deficit. Yet, the various administrations could not make this transition due to political reasons. Basically, the inward oriented Turkish industrial products were unable to compete in the international markets on product quality and the producer companies were insisting on continued government protection. Likewise, the agricultural sector did not provide the needed export earnings. Simply, the traditional levels of agricultural exports, at an average of $1.29 billion annually between 1976 and 1979, were not sufficient to meet Turkey's import needs. Thus, by the end of the 1970s, Turkey's export earnings stagnated while imports continued to rise rapidly under conditions of high domestic demand and worsening terms of trade.

To combat these developments, the nationalist coalition in 1978 and the RPP administration in 1979 undertook two IMF-based attempts at economic stabilization. They implemented several devaluations of the national currency and tried to enforce tight monetary policies. Both measures failed to improve the trade deficit basically due to the increasing political crisis in the country. Political violence burdened the administrations' ability to complete emergency economic policies to their necessary levels. In fact in October 1979, the IMF survey of the Turkish economy found that the tight monetary policies had been heavily relaxed and "stand-by" agreements violated (OECD, 1980: 25). In an attempt to reverse these developments, the JP administration adopted an IMF based economic austerity package on 24 January 1980. This package, known as the January Measures, included the following key elements (OECD, 1980: 32-35):

1. Institutional changes aimed at making policy formulation and implementation more effective;
2. Devaluation of the Turkish Lira vis-à-vis the U.S. dollars by 33 percent and the limitation of multiple exchange rate policies;

3. Greater liberalization of the trade regime and payment regime;
4. Additional promotional measures for exports;
5. Substantial price increases for state traded goods and abolition of price controls;
6. Increased competition for State Economic Enterprises and abolition of most government subsidies;
7. Higher rates of interest;
8. Promotion of foreign investments;
9. Arrangements for consolidating Turkey's private commercial debt;
10. Draft legislation for tax reforms;
11. Reduce domestic consumption level.

In essence, these measures called for dramatic changes in Turkey's macroeconomic growth strategies. For the first time during the Second Republic, a ruling elite coalition embarked on a new policy designed not only as a serious comprehensive program of economic stabilization but also as a basic reorientation of economic policy away from detailed government regulation and control toward greater reliance on market forces, foreign competition, and foreign investment. That is, a new chapter opened in economic development that emphasized export-orientation to integrate Turkey's economy with the world market. Application of these policies in the ensuing years profoundly affected the economy. For instance, the state subsidies, price controls and trade regulations that were provided as support for the agricultural sector during the pre-1982 years were now to be eliminated.

DEVELOPMENTS IN THE AGRICULTURAL SECTOR

Agriculture Price Support Programs

State agricultural price supports largely protected domestic producers against foreign competition. Originally this program began in 1932 when the state decided to protect wheat producers against the world-scale decline in wheat prices caused by the great depression. In the following years, partly due to central economic planning and partly to their electoral implications, agricultural price supports became a standard policy of Turkish governments. Table 6.1 provides the recent figures on support prices for various agricultural commodities. These figures represent prices set by the state for the purchase of these items directly from the producers.

Table 6.1
Agricultural support prices (real TL/kg)[a]

Commodity	1967	1972	1977	1978	1979	1980	1981	1982
wheat	0.62	0.46	0.53	0.37	0.36	0.39	0.49	0.82
barley	0.34	0.37	0.37	0.56	0.33	0.32	0.37	0.31
rye	0.44	0.35	0.35	0.28	0.33	0.31	0.34	0.30
cotton	1.68	1.75	1.99	1.77	1.74	1.80	1.64	1.51
tobacco	6.22	5.47	8.20	5.72	4.29	4.03	3.39	4.16
hazelnuts	4.05	3.98	3.05	2.62	3.14	3.95	3.26	2.95
tea	2.70	1.87	1.85	1.37	1.01	0.94	1.07	1.08
sugarbeets	0.11	0.09	0.11	0.10	0.10	0.11	0.12	0.10
olive oil	3.94	4.07	4.25	4.05	4.19	4.49	--	--
raisins/seedless	1.75	1.37	2.22	2.68	3.14	3.05	2.87	2.60
raisins	--	1.17	1.39	1.14	1.40	1.44	--	--
dry figs	1.16	1.22	1.48	1.43	1.54	1.80	1.70	1.53
pistachios	--	6.08	4.90	6.28	10.47	10.78	--	--
sunflower seeds	--	1.03	1.20	0.97	1.12	1.08	1.04	0.98
mohair	--	14.88	21.25	21.68	25.82	13.23	20.88	--
wool	--	13.03	10.16	8.21	9.77	10.45	8.48	--
live animals	--	--	4.84	4.39	3.30	--	--	--
rose flower	--	--	--	1.26	0.94	1.08	--	--
poppy	--	--	--	2.66	1.72	1.22	0.98	--
lentil	--	--	--	--	1.19	1.01	--	--
groundnuts	--	--	--	2.11	1.95	2.51	1.83	1.39
silk cocoon	--	--	18.47	14.26	12.91	28.73	--	--

[a] Current figures are converted to real TLs by using the consumer price index with 1963=100.

Source: Minisrty of Finance, Economic Report 1971, p. 88 and Economic Report 1982, p.127; Ministry of Finance, Annual Report 1981, p. 142; and SPO, IV. Five Year Development Plan, p. 123.

The figures in Table 6.1 suggest significant developments in this area. The most apparent development over the last twenty years is the substantial increase in the number of commodities subjected to price supports. Until 1963, this program only covered eleven commodities (Kazgan, 1977: 305). During the Planned Period, this number increased to twenty-two. As stated above, price supports protected the producer's income levels against fluctuations in the prices of their commodities in world markets. Additionally, the administrations attempted to eliminate the gap between foreign and domestic market prices and to maintain the production of certain commodities at pre-determined levels. However, the real figures in Table 6.1 show that between 1967-1982 support prices for most commodities have actually declined or at best remained constant.

Besides economic concerns, political factors also have affected price support levels at different times. These political factors have been both domestic and external, with both positive and negative implications for the producers. U.S. pressure on Turkey to ban poppy production in 1971 demonstrates one example of an external factor. Following the Turkish government's decision to prohibit poppy production, the U.S. administration agreed to pay Turkey $35 million compensation over a three-year period. However, the program proved so damaging to the producers that the RPP-NAP coalition in power during 1974 immediately moved to permit state controlled production of poppy. Simply stated, the $35 million in assistance did not even approximate producer's income for poppy and most farmers had difficulty shifting production to alternate crops. Similarly, as discussed later, the IMF-sponsored austerity measures in 1980 resulted in reduced state subsidies in dollar amounts.

Domestic political factors largely relate to the election considerations of political parties. The RPP-NSP coalition of 1974 and the RPP-independents coalition of 1978-October 1979 increased price support levels. During these periods, the RPP attempted to increase its rural voter support and to change the party's image of being a urban-based social democratic institution.

An expansion of funds in the Central Bank allocated for price supports during the Planned Period allowed an increase in price support practices. The Central Bank credits used for price supports were: $1,180 million TL in 1963, $3.314 million TL in 1967 (Kepenek, 1983: 334-335). However, the proportion of these figures to total credit of the Central Bank was largely the same: 37.45 percent in 1963; 37.76

percent in 1967; 38.65 percent in 1972; and 38.18 percent in 1976. That is, while the monetary value of credit for price supports had increased between 1963 and 1976, their real value within the total credit budget of the Central Bank remained relatively constant. In this sense the agricultural sector's receipt of funds from the expanding national budget remained constant.

Following the adoption of the IMF-sponsored export-oriented economic package in 1980, price support programs began to suffer. As shown in Table 6.1, the overall actual price paid for most commodities shows a decline since 1980. Furthermore, the total amount paid to the producers shows slower growth rates for each year. While in 1980 the growth rate (in current value) was 98.4 percent, this figure sank to 42 percent in 1981 and further declined to 26 percent in 1982 (SPO, 1982: 90; SPO, 1983: 113). The declines in price support programs under the new economic policy also shows in the public funds allocated for this program. Since 1980, the proportion of funds allocated for price supports to total Central Bank credits declined significantly, reaching an all time low of 18.1 percent in 1982. That is, one major outcome of the government's rightist monetary policies since 1980 has been the weakening of state subsidies for the agricultural sector.

An additional problem has been the constant delay in governments' payment for state-purchased goods, which has created difficulties for agricultural producers in Turkey. Since 1979, producers have been waiting up to one year for government payments for the goods purchased under the price support programs. This practice causes the constant deterioration of the producer's purchasing power in the market.

In conclusion, during the import-substitution growth phase, price support programs contributed to the maintenance of a state-supported strong agricultural sector in the economy. The practice was not perfect, but nevertheless, it did not reduce the agricultural sector's receipt of funds from the growing national budget. Since the adoption of the IMF austerity measures, the agricultural sector does not receive similar state attention. The 1980 coup agreement and the present Özal administration have abided by the IMF measures which call for reduced government subsidies in the national economy. As a result, rural producers can no longer count on adequate state support for their products.

Modernization of Agricultural Production

In modern agriculture, increase in productivity is associated with the use of modern inputs for production. These inputs of the last three decades include farm machinery, fertilizers, chemicals and high yield seeds and are called "Green Revolution" technology. The intensified use of the Green Revolution inputs in Turkey began in the early 1960s and have continued to the present. Table 6.2 provides the data on the use of Green Revolution technology in Turkish agriculture.

The data indicate that by the early 1960s the area of land under cultivation was near its maximum size. Between 1962 and 1980 the area of cultivated land varied between 23,719,000 - 24,972,000 hectares. This indicated that increases in agricultural production during the Planned Period cannot be attributed to the opening of new areas for cultivation. Rather, changes in the nature of production processes caused increases in productivity; the intensified use of tractors, chemicals and fertilizers characterize these changes.

It is interesting to note that during the Planned Period the use of fertilizers and chemicals received the most attention. Between 1962 and 1967, the use of fertilizers and chemicals per year increased to 39.6 percent. During the Second and Third Plan periods (1969-1977), this increase was around 17.4 percent annually. However, these figures could be misleading because despite the dramatic increase in use of fertilizers and chemicals between 1963 and 1967, the area subjected to these new inputs remained around 25 percent of total agricultural land. Furthermore, at the end of the Third Plan in 1978, this proportion increased to 47.2 percent and in 1982 it leveled off at 55.0 percent (Kazgan, 1983: 125-127). The size of farms in Turkey can partially account for this development. As shown later, the majority of Turkish farms (82.0 percent of total farms in 1980) are small holding farms with less than 10 hectares. It is expensive for these farmers, who often rely on subsistence farming, to use new fertilizers and chemicals.

The use of tractors in agriculture shows a significant increase during the Third Plan. During this period, the number of tractors in use increased from 156,129 in 1973 to 320,578 in 1977 at an average rate of 18.9 percent annually. On the average, the increase in the number of tractors during the first two Plans was 11.6 percent and 12.6 percent respectively. In 1982, this number increased to 491,000, bringing the percentage of land under tractor use to around

Table 6.2
Indicators of agricultural modernization

Year/Period	Cultivated Area (000 hectares)	Tractors number	Tractors % increase	Use of Chemicals and Fertilizers (000 tons)	Use of Chemicals and Fertilizers % increase	Agricultural Credits TL (million)	Agricultural Credits % increase
1962-1967	23,719	54,854	11.6	770.3	39.5	3,263.5	22.0
1968-1972	24,531	108,460	12.6	2,520.6	17.5	7,151.6	16.0
1973-1977	24,558	240,410	18.9	4,613.9	17.4	32,612.7	39.1
1978	24,552	370,259	15.5	7,474.0	13.6	64,177.9	34.5
1979	24,972	402,777	8.8	7,666.0	10.3	119,643.4	86.4
1980	24,568	436,369	8.3	5,967.5	-22.2	184,957.6	54.5

Source: State Statistics Institute (SSI), Statistics Yearbook, 1968, p. 165, 168, and 180; Statistics Yearbook, 1973, p. 188 and 201; Statistics Yearbook, 1977, p. 165 and 175; and Statistics Yearbook, 1981, p. 176-177, 186, and 188.

79.2 percent. By this time, even some of the smaller farms in Turkey were utilizing tractors in their agricultural production. The figures on agricultural credits can be helpful in explaining this development. Experts have noted that the Turkish administrations have subsidized the cost of Green Revolution inputs for farmers and have also provided low interest agricultural credits for the purchase of necessary goods (Kazgan, 1983; Kepenek, 1983: 321). In general, more farmers have used their credits to purchase machinery than those who have also obtained fertilizers and chemicals. This is especially the case for the small farmers.

Agricultural Productivity, Exports, and Food Supply

The increased use of modern inputs in agriculture during the Plan Period resulted in higher yields in agricultural commodities. Table 6.3 provides index figures on commodity production. The commodities are classified with four groups: cereals, fruits and vegetables, industrial crops, and live animals. Industrial crops include tobacco, cotton, oil seeds, tea, and sugarbeets.

The figures show that during this time period, agricultural yield improved substantially. Cereal production increased from an index figure of 71.4 in 1963 to an average of 105.7 in early 1980s. Similar growth is also apparent in the production of fruits and vegetables and industrial crops. The yield in fruits and vegetables increased from 66.6 in 1963 to over 119.9 in 1982. Likewise, the yield in industrial crops increased from an index of 39.1 to an average of 112.0 in the early 1980s. Undoubtedly, such dramatic growth in production of agricultural commodities resulted from increased use of modern inputs in agriculture, since as shown previously, there had been little increase in the size of total cultivated area in Turkey between 1963 and 1983.

Animal husbandry also shows growth during this period. The total number of recorded live animals (cattle, sheep, goats, poultry) was 86.1 million in 1963. This figure increased steadily over the next twenty years to reach 146.1 million in 1983. An analysis of these figures with regard to Turkey's macroeconomic growth policies produces significant results. During the import-substitution period (1963-1979), cereal production, based on index figure, increased at an average annual growth rate of 8.6 percent. In the same period, the increases for fruits and vegetables, industrial crops, and live animals were 2.9, 8.7, and 5.1 percent respectively. Interestingly enough, during the export-led growth period of 1980-1983, cereals production decreased by an average annual rate of -0.9 percent. The growth rate of fruits and vegetables and live animals dropped down to an average annual figure of 1.0 percent. All of these figures are before the average annual growth rates of 1963-1979 for these commodities. The only group of agricultural commodities that showed a greater annual growth rate during the export-led period are the industrial crops. This group's average annual growth rate jumped from 5.1 percent during 1963 to 1979 to 10.5 percent during 1980 to 1983. This suggests that during the export-led growth period the emphasis on agricultural productivity somehow shifted to favor industrial crops over the traditional commodities.

Table 6.4 provides figures on exports of agricultural commodities during 1975 to 1983. Reliable commodity group statistics do not exist for the earlier years to enhance the comparison with production figures. Nevertheless, we can compare the latter years of import-substitution phase and the early years of the export-growth model.

During 1975 to 1979 the value of cereals exports increased from $793 million U.S. dollars to $1,344 million, or increased from an index figure of 63 to 107, at an average annual growth rate of 13 percent. Under the export-led growth program, export of cereals shows an average annual growth rate of 10 percent between 1980-1983. This reduction in the growth rate of cereals exports is in accordance with the observed reduction in their production rate. That is, as the growth rate of cereals decreased under the export-led model, so did their exports. Similar declines are observed for the annual export growth rates of fruits and vegetables. During 1975 to 1979, exports of fruits and vegetables increased at an average annual rate of 24 percent. In comparison, this figure was -1.0 percent between 1980 and 1983. These observations confirm the above argument that the two traditional export commodity groups have lost their predominant role in the export scene. For cash crops, on the other hand, the post-1980 exports show an average annual increase of 15 percent as opposed to their 10 percent average growth during 1975-1979.

One commodity group where decline in annual productivity rate was not followed by similar decline in export growth rate is live animals. The average annual growth rate of export of live animals increased from 23.0 percent between 1976 and 1979 to 52.0 percent during 1980 through 1983.

Table 6.3
Agricultural production figures (index 1976=100)

Commodity group	1963	1967	1972	1976	1979	1981	1982	1983
cereals	71.4	65.1	76.5	100	104.8	104.3	108.9	103.6
fruits and vegetables	66.6	96.9	101.8	100	114.7	123.1	119.9	120.8
industrial/ cash crops[a]	39.1	59.1	69.8	100	94.3	113.3	128.4	149.2
live animals	69.8	84.5	89.1	100	113.7	117.0	116.8	118.4

[a] includes tobacco, cotton, oil seeds, tea, and sugarbeets

Source: SSI and SPO archive data.

Table 6.4
Index of agricultural exports (1976=100)

Commodity group	1975	1976	1977	1979	1981	1982	1983
cereals	63	100	83	107	177	171	150
fruits and vegetables	74	100	117	150	212	173	157
cash crops[a]	59	100	89	103	211	281	342
live animals	73	100	65	112	330	552	509

[a] including processed goods like cigarettes.

Source: SPO archive data.

This is a significant increase in the value of live animals. The average annual growth rate of live animals declined to 1.0 percent in the early 1980s and the official export support prices of live animals have remained around 47.29 TL to 65.30 TL per kilo over this period (Kepenek, 1983: 338 and SPO, 1985: 140-142). The current Iran-Iraq war can explain this sudden jump in the export of live animals since 1980. Since this war began in 1980, both Iran and Iraq steadily increased their imports from Turkey. In 1980, only 7.0 percent of Turkey's total exports went to Iran and Iraq (SPO, 1985b: 27). This figure increased substantially to 24.4 percent in 1983 and since has remained around 25 percent (SPO, 1986). The official figures on trade with Iran and Iraq do not necessarily show the significance of this relationship because there is another side to trade relations between these commodities. Part of the Turkish trade between Iran and Iraq takes place in the form of barter trade where Turkish goods, ranging from food stuff to weapons, are exchanged for petroleum. We have no way of measuring the exact dollar amounts of such transactions, although Turkish officials acknowledge its existence. In any case, the Iran-Iraq war is a convenient channel for the Turkish authorities to increase the countries export earnings.

One additional factor to report is the increases in exports of certain agricultural commodity groups and their association with Turkey's external debt obligations. Following the adoption of the IMF measures in January 1980, a time frame was drafted for Turkey's repayments of a $15.76 billion external debt. In fact, the main argument behind reduction of domestic consumer demands through the reduction of real wages and emphasis on export earnings was to enable Turkey to meet her external debt obligations. The original plan (Ministry of Finance, <u>Information Memorandum</u>, 1983) called for annual repayments of $1.93 billion in 1983, $2.05 billion in 1984, $2.53 billion in 1985, $2.39 billion in 1986, $2.43 billion in 1987, $2.41 billion in 1988, $2.07 billion in 1989, and $1.52 billion in 1990. However, as a direct result of the post-1980 administrations' reliance on further borrowing from abroad, Turkey's total foreign debt increased to $19 billion in December 1983 and continued to reach the $30 billion mark in January 1987 (<u>Wall Street Journal</u>, 7 January 1987). The steady growth of Turkey's foreign debt resulted in revisions in her debt repayment time frame. The new time frame adopted in 1986 places more demands on the export performance of the economy. According to this new time frame, Turkey's annual debt repayments are

$4.36 billion in 1986, $4.49 billion in 1987, $4.61 billion in 1988, $4.16 billion in 1989 and $3.57 billion in 1990 (Maliye Bakanligi, 1986). With such dramatic increases in debt repayments from the 1983 agreement, Turkey's debt service ratio jumped from 6.3 percent in 1980 to 45 percent in 1985 (Cumhuriyet, 11 November 1986). In other words, 45 percent of Turkey's export earnings went to service her foreign debt obligations in 1985. The unofficial figure for 1986 was a record high of 57 percent.

Under such debt obligations, the Turkish officials desire to increase the country's export earnings is understandable. This is one reason why the production and export of industrial crops has risen since 1980. Yet, the price Turkey is paying for this policy seems to be the weakening of the domestic food base as the export rates of traditional crops are greater than their production rates. As shown above, the animal husbandry sector of the economy best describes this.

These increases in the export of live animals, without comparable growth rates for their replacement, undoubtedly threatens the domestic food base of Turkey. Until recent years, domestic food supply was not a serious political issue. Until the early 1980s, Turkish farmers produced food and other agriculture commodities at a rate that seemed sufficient for domestic demand and export needs. During 1963 through 1979, the average per capita annual growth rate of traditional food commodities was 4.0 percent while the average population growth rate was 2.5 percent (FAO, 1968-1980). Examining the growth of per capita income of about 3.9 percent per year and an income elasticity of demand for food of 0.3 (World Bank, 1980: 101) shows that during the first three Five Year Development Plan period, the growth rate of the production of cereals, fruits and vegetables, and live animals sufficiently met the growth rate of demand for food. Data in Table 6.5 on wages and consumer price indices also support this observation.

These figures show that with each successive Plan Period the index of prices substantially increased. While the average annual increase for the consumer price index was 6.8 percent in 1963-1967, it jumped to an average annual figure of 20.5 percent during 1973-1977. However, in the same years the average annual increases in real wages were higher than those of the consumer price index. This trend continued until 1978 and holds for the industrial workers and bureaucrats. Since the data on the wages of agricultural workers are largely lacking during this period, it is not possible to make a similar comparison. These figures

Table 6.5
Average annual change in daily wages of workers[a] (percent)
(wages are given in TL)

Period	Consumer Price Index (1963=100)		Average Daily Wages of Workers			
	Total	Food	Agriculture	Industries		
				Public	Private	Bureaucracy
1963-1967	6.8	7.5	--	14.0	8.1	7.9
1968-1972	10.6	9.9	--	13.5	13.8	13.5
1973-1977	20.5	21.6	--	32.5	25.1	25.5
1978-1980	73.2	66.3	57.4	67.5	56.6	44.9
1981-1984	37.8	36.8	20.5	29.7	27.7	19.2

[a] real wages are obtained by adjusting the current wages by the consumer price index.

Source: SPO archive data.

suggest that until 1978 the average Turkish bureaucrat and industrial worker did not face serious difficulty in meeting his basic food needs. It should be noted, however, that this generalization is not valid for the urban poor or the landless peasants. We have no reliable information on estimates of suffering by these two groups during this period. With the beginning of the Fourth Five Year Development Plan the above observation for the industrial workers and bureaucrats reversed and took a turn for the worse following the adoption of the 1980 IMF austerity measures.

Since the adoption of the export-oriented growth model, domestic food supply and the consumers basic ability to afford these goods has worsened. On the one hand, annual growth rates of basic food commodities have declined to a level that is below the population growth rate. Between 1980 and 1983, the average per capita annual growth rate of basic food commodities was 2.1 percent, while the average population growth rate was 2.3 percent (FAO, 1985). On the other hand, as a direct result of the governments' tight monetary policies aimed at curbing domestic consumer demands (dictated in the January 1980 IMF measures), annual wage increases in all sectors of the economy have fallen behind the annual price increases for food commodities. During 1981 through 1983, the price index of food commodities increased at an average annual rate of 36.8 percent. In retrospect, the average annual increase of wages for this period were: agriculture, 20.57 percent; private sector industries, 27.7 percent; public sector industries, 29.7 percent; and bureaucracy, 19.2 percent. In other words, wages remained 7-17 percent behind price hikes for food. As a result, wage earners in Turkey face serious difficulties in meeting their household's subsistence needs. The research of concerned scholars and journalists (Cumhuriyet, various issues) highlights the seriousness of this problem for Turkey's urban dwellers. The researchers have tabulated the basic food needs of a family of four (average for middle-income urban class) based on market prices. The monthly bill for these items was 29,715 TL in early 1983 and has since increased at an alarming rate. The increases are: 37,600 TL in January 1984, 58,320 in January 1985, 71,645 TL in April 1985, and 94,390 TL in June 1986. In comparison an average monthly salary (net) of an assistant professor in Turkey in 1985 was 120,000 TL. Considering the family's other subsistence needs and an average rent for apartments in major urban cities (50,000-150,000 TL in one month), even the upper middle-class struggles to meet their everyday needs. Of course, lower wage earners suffer more. The gross nonagricultural minimum wage in Turkey was a mere 10,000 TL in 1982 and 41,400 TL in 1985. In net wages, these figures translate to 7,125 TL and 27,812 TL respectively. The gross minimum wages in agriculture for these years were 8,610 TL and 34,000 TL (SPO, archive data).

In summary, export-oriented growth in Turkey has favored the production of cash crops over traditional commodities. At the same time, state control of wage increases has resulted in a serious crisis for average citizens in meeting their basic subsistance needs.

PROBLEM OF LAND DISTRIBUTION

One additional factor inhibits agricultural development in Turkey---the distribution of land among farmers. This crucially affects Turkey's efforts toward economic development since the agricultural sector still maintains a substantial role in determining the level of exports and GDP, and employs the majority of Turkish workers. Table 6.6 provides data on land distribution in Turkey. It should be noted that the latest reliable figures on land distribution are of 1973.

These figures demonstrate that land inequality in Turkey slightly worsened during 1963 through 1973. The Gini coefficient during this period increased from 0.59 to 0.63. The ratio of farmers who owned 0.1 to 2.0 hectares of land increased from 40.7 percent to 44.6 percent. However, their total land owning declined from 11.3 percent to 8.4 percent in the same years. On the other hand, the ratio of the largest landowners, with 50 plus hectares, increased from 0.5 percent in 1963 to 0.8 percent in 1973. But unlike the smallest farm owners, the land holdings of the largest farmers increased from 10.7 percent to 14.1 percent in the same years, respectively. By 1973, small farmers (with 0.1-5.0 hectares) comprised 72.9 percent of total farm households. Their share of land, however, accounted for a mere 26.2 percent.

In addition to the disproportionate distribution of land among farmers, the State Planning Organization (1973: 13) reported that 21.9 percent of rural households held no land. This figure was 31 percent in 1963 and 17.5 percent in 1968. That is, during the first two Five Year Development Plans, the number of landless rural households more than doubled. According to Varlier (1978: 3), the largest concentration of landless peasants was in the South and Southwestern regions

Table 6.6
Land distribution in Turkey

Unit Size (hectares)	1963 (percent) Household	1963 (percent) Land	1973 (percent) Household	1973 (percent) Land
0.1 - 2.0	40.7	11.3	44.6	8.4
2.1 - 5.0	28.1	17.7	28.3	18.2
5.1 - 10.0	18.1	22.2	16.7	23.0
10.1 - 20.0	9.4	22.2	7.0	19.8
20.1 - 50.0	3.2	15.9	2.6	16.6
50.1 - over	0.5	10.7	0.8	14.1
Total (%)	100.0	100.0	100.0	100.0
Total (n)	2,965,203	15,961,806(ha)	2,965,446	16,842,810(ha)
Gini Coefficient	0.59		0.63	

Source: SPO, IV. Five Year Development Plan (Ankara, 1979), p.12.

of the country with the lowest concentration centered in the Black Sea provinces.

Several factors caused the poor state of land distribution in Turkey. First, inheritance customs often result in land being divided among the surviving sons of a farmer. Second, and more significant for the time period considered, is the intensified application of Green Revolution technology in Turkey during the first two Five Year Development Plans (1963-1972). Green Revolution technology inputs (fertilizers, pesticides, improved high yield seeds, and machines) are costly and, as in many other developing countries, mainly large land owners, use them. Furthermore, as productivity in these farms increased, their market competitiveness also improved resulting in the concentration of more land in the hands of large land owners who had 20.9-50+ hectares. A third factor aggravated this problem; lack of land reform has persisted since the establishment of the Turkish Republic in 1923.

During the First Five Year Development Industrialization Plan (1933-1937), the state helped rapid industrial growth but failed to undertake similar measures in agriculture (Inan, 1972). Throughout the 1930s and 1940s the RPP leaders emphasized their devotion to the interests of the peasants; this class had occupied the lowest citizenship status under the Ottoman Empire. Yet, since the etatists relied on the support of the rural land lords, known as "ağa"s, for their political control of the countryside, the RPP could not adopt any land reform measures that might alienate this group. As a result of the alignment of urban RPP elites and rural "ağa"s, land reform did not accompany industrialization efforts. Instead of real land reform that would eradicate the unequal distribution of lands, the state distributed some of the land under its control to landless peasants and migrants from the Balkans. Between 1923 and 1924, the state distributed some 711,322.5 hectares of land. In comparison, this figure was 299,892.5 hectares between 1934 and 1938. The combined figures added up to about 7.6 percent of the total cultivated land in Turkey. The average distribution of land was: a Balkan family, 4 hectares; landless peasant 31 hectares; and tribal family, 1.7 hectares (Barkan, 1944). Needless to say, such minimal state gestures towards the peasants and Balkan migrants did not alarm the local "ağa"s. After all, this move left the large private holding untouched.

However, in 1945 the étatists drafted the Law of Granting Land to the Farmer and Establishing Farmers Cooperatives (No. 4753) and proposed it to the National

Assembly. This new law explicitly stated that land could be granted to landless peasants by expropriation from state owned lands, evkafs, and private estates of over 500 hectares. If this proved insufficient, the figure could be lowered to include private properties over 200 hectares (Hale, 1981: 62-63). "Ağa"s and prominent RPP deputies in the Assembly who were also large landowners strongly criticized this bill. Some of these deputies broke away from the RPP in 1946 and formed the opposition Democrat Party, which attracted the political support of the bourgeoisie and "ağa"s in Turkey and came to power in the first democratic elections in 1950. During 1946 through 1950, even though the Assembly approved Law No. 4753, the politically weakened RPP was unable to carry out its principles. The total transfer of land amounted to 162,000 hectares by 1951. Once the Democrat Party came to power, its members viewed the land reform law as communist policy and overturned it. This decision ended the first attempts at achieving land reform in Turkey but the idea itself plagued Turkish politics for years to come.

During the next ten years of Democrat Party rule, the American tractors arrived in Turkey under the Marshall Plan. The Americans had an earlier interest in promotion of agricultural development in Turkey as apparent in the U.S. economic missions' reports on the Turkish economy (Thornburg, 1944; IBRD, 1951: 32-33, 57). Both reports emphasized the need for Turkey to abandon rapid industrialization and to produce a capitalist model of development in agriculture. However, the reign of "ağa"s in the rural sector did not end. Rather, the large landowners adopted American agricultural technology through favorable state credits for improved and efficient use of the land. The grand design behind this American initiative was to establish Turkey as the bread basket of Europe in conjunction with the postwar recovery program. With the arrival of American tractors, the agricultural sector grew rapidly during 1951 through 1955. Yet, mismanagement of the economy by the Democrats, poor harvests in the late 1950s, and growing social unrest led to the military coup of 27 May 1960.

The 1960 coup marked a new phase in Turkish history. The officers adopted a new constitution that included the need for planned economic growth. Land reform has not gained much ground, despite a clear statement in the 1961 Constitution that the state should adopt measures needed to achieve the efficient utilization of land and provide land to those farmers who either had or owned insufficient land

(Republic of Turkey, 1961: Arts 37-38). The 1961-1970 civilian and 1971-1973 military-based agreements made no attempts to carry out land reform in Turkey, largely due to the political clout of "ağa"s in the powerful center-rightist Justice Party. Finally in 1974, the coalition government of the Social Democratic RPP and the fundamentalist NSP passed and attempted to implement the 1973 Land Reform Law.

According to this law, some 3.2 million hectares were to be distributed to the 540,000 peasant families that made up 65 percent of the total number of landless peasants in 1973. In villages where land reform was to be applied, the law also called for the establishment of peasant cooperatives, known as Agrarian Reform Cooperatives, to provide tools, credit, and marketing facilities to the farmers (Toprak ve Tarım Reformu Kanunu, 1973: Arts 7, 70-71, 81-83). Implementation of this law began with a pilot project in Urfa where there was a large concentration of landless peasants working for "ağa"s. The government appointed Professor Saim Kendir to distribute 300,000 hectares in Urfa region. After several months, Professor Kendir reported that the local landlords had established a secret fund to prevent the implementation of land reform and to threaten the peasant who were approaching government officials for help (Avcıoğlu, 1981: vol. 2, 699-670). When these efforts were interrupted by the 1974 war in Cyprus, and later the resignation of the coalition government, land reform in Urfa came to a sudden halt. The succeeding Nationalist-Front coalitions, led by the Justice Party, did very little to complete the plan.

By 1977, the Constitutional court annulled the entire project on procedural grounds and only 23,000 hectares had been distributed to the landless peasants. Since then, no real effort has been made, even after the 1980 coup, to carry out land reform in Turkey. In essence, the future of this crucial and sensitive issue seems quite bleak given the fact that the post-1980 administrations consider the topic "taboo". In sum, the failure of real land reform and the concentration of new agricultural technology in the hands of large and middle size farmers have both contributed to the worsening of land distribution in the country. As indicated above, the number of landless peasant families in Turkey more than doubled during the first Five Year Development Plan Periods. Officials in the SPO indicate that the figures have become worse in recent years. The State's ineffectiveness in carying out land reform or protecting small farmers against the large land owners have both contributed to this end.

CONCLUSION

The above analysis has demonstrated that during 1963 through 1978 Turkish agricultural development progressed steadily to meet two essential requirements of the state. The first requirement was the production of sufficient food supply to meet domestic needs. In this regard, no major problem existed for the producers to meet domestic market needs. At the same time, the consumers were by and large able to afford their food needs as their wages increased steadily. The second requirement was the need to increase export earnings of the state through sales of cereals, fruits and vegetables. These requirements were carried out under the import-substitution development policies of the ruling government coalitions, both civilian and military alike. Various price support programs and credits for the purchase of modern agricultural inputs were extended to the farmers to protect their commodities in world markets and to ensure voter support for ruling political parties in future elections. Thus in aggregate terms, Turkish agricultural development during 1963 through 1978 succeeded. However, in microeconomic terms the picture was less successful as the degree of land inequality among Turkish farmers worsened. The gini index for land distribution increased to 0.63 in 1973 as intensification of the Green Revolution technology favored large landowners at the expense of small farmers.

With the adoption of the January 1980 economic package, Turkish agricultural development became attached to the country's foreign debt obligations. On the domestic front, the tight monetary policies of the IMF resulted in the serious depreciation of the real wages of Turkish workers because of the maintenance of current wage increases below the level of the annual inflation rate. As real wages declined, another goal of the IMF austerity measures was also achieved---reduction in the domestic consumption level.

In terms of Turkey's external needs these IMF policies emphasized increased exports from all sectors of the economy to finance the country's foreign debt repayments. The agricultural sector responded to this call by increased exports of cash crops (industrial crops) and live animals. This emphasis on agricultural commodity exports coupled with the fact that between 1980 and 1983 the average per capita food production fell below the population growth rate shows that the vast majority of Turkish citizens have difficulty in meeting their basic subsistance needs.

Serious inequality in land distribution further aggravates this situation among Turkish farmers. As the Turkish economy becomes more integrated with the world market and economic growth is left to market forces via removal of state subsidies and protectionist trade measures, more and more small farmers are unable to compete with large-scale commercial enterprises. As long as the Turkish governments refuse to acknowledge the seriousness of this matter, land inequality in Turkey and the reign of large landlords will continue to remain as thorns in the country's efforts at agricultural development.

REFERENCES

Adalet Partisi (1973). Adalet Partisi Seçim Programı [The Election Program of the Justice Party]. Ankara.

_____ (1977). Adalet Partisi Seçim Programı [The Election Program of the Justice Party]. Ankara.

Avcıoğlu, Doğan (1981). Türkiyenin Düzeni [Turkey's Order] Istanbul: Remzi Kitabevi.

Barkan, Ömer (1944). "Laloi sur la Distribution de Terres aux Agriculteurs et les Problems Essential d'une Reforme Agraire en Turqui." Revue de la Faculte des Sciences Economiques de l'Universitee d' Istanbul.

Cumhuriyet (Istanbul). Selected Issues.

Cumhuriyet Halk Partisi (1976). CHP Seçim Programı [The Election Program of the RPP]. Ankara.

Ergil, Dogu (1980). "Electoral Issues: Turkey, in Electoral Politics in the Middle East: Issues, Voters, and Elites eds. E. Ozbudun, F. Tachau, and J.M. Landau. London: Croom Helm.

FAO (1968-1980). Production Yearbook. Rome: FAO Publications.

Hale, William (1981). <u>The Political and Economic Development of Turkey</u>. New York: St. Martin's Press.

İnan, Afet (1972). <u>Devletçilik İlkesi ve Türkiye Cumhuriyeti Birinci Sanayi Planı</u> [The Principle of Estatism and the Turkish Republic's First Industrial Plan]. Ankara: Türk Tarih Kurumu Yayınları.

IBRD (1951). <u>The Economy of Turkey</u>.
Baltimore: John's Hopkins University Press.

Kazgan, Gülten (1977). <u>Tarım ve Gelişme</u> [Agriculture and Development]. Istanbul: Istanbul University Faculty of Economics Press.

_____ (1983). <u>Tarım ve Gelişme</u> [Agriculture and Development] 2nd. ed. Istanbul: Der Yayınları.

Kepenek, Yakup (1983). <u>Türkiye Ekonomisi</u> [The Turkish Economy]. Ankara: ODTÜ Yayınları.

Kongar, Emre (1981). <u>Türkiyenin Toplumsal Yapısı</u> [The Social Structure of Turkey. Istanbul: Remzi Kitabevi.

Ministry of Finance (1971). <u>Economic Report</u>.
Ankara: Ministry of Finance.

_____ (1981). <u>Annual Report</u>.
Ankara: Ministry of Finance.

_____(1982). <u>Economic Report</u>.
Ankara: Ministry of Finance.

_____(1983). <u>Economic Memorandum</u>.
Ankara: Ministry of Finance.

_____(1986). <u>Economic Indicators</u>.
Ankara: Ministry of Finance.

OECD (1980). <u>Economic Surveys: Turkey</u>.
Paris: OECD Publications.

Özubudun, Ergun (1980). <u>Electoral Politics in the Middle East: Issues, Voters, and Elites</u>. London: Croom Helm.

Republic of Turkey (1961). <u>The Constitution of The Republic of Turkey</u>. Ankara.

SPO (1969). Second Five Year Development Plan. Ankara.

_____ (1973). Third Five Year Development Plan. Ankara.

_____ (1979). Fourth Five Year Development Plan. Ankara.

_____ (1980). Economic Program, 1980. Ankara.

_____ (1982). Economic Program, 1982. Ankara.

_____ (1983). Economic Program, 1983. Ankara.

_____ (1985). V. Beş Yıllık Kalkınma Planı Öncesinde Gelişmeler 1972-1983 (Ekonomik Gelişmeler) [Economic Developments Prior to the Fifth Five Year Development Plan]. Ankara: SPO Publications.

_____ (1985b). Main Economic Indicators. Ankara.

_____ (1986). Main Economic Indicators. Ankara.

State Statistics Institute of Turkey (1968, 1973, 1977, 1981). Statistical Yearbook. Ankara.

Thornburg, M.W. (1949). Turkey: An Economic Appraisal. New York: Greenwood.

Tokgöz, Erdinç (1976). Sanayileşmede Bölgesel Dengesizlik ve Türkiye [Turkey and the Regional Unbalance in Industrialization]. Ankara: Hacettepe University Press.

Toprak ve Tarım Reformu Kanunu (1973) [The Land Agriculture Reform Law of 1973]. Ankara.

Varlıer, Oktay (1978). Türkiye Tarımında Yapısal Değişme, Teknoloji, ve Toprak Bölüşümü [Structural Change, Technology, and Land Distribution in Turkish Agriculture]. Ankara: SPO Publications.

Wall Street Journal. 7 January 1987.

World Bank (1980). <u>Turkey: Policies and Prospects for Growth</u>. Washington, D.C.: IBRD Publications.

Yeşilada, Birol A. (1984). <u>Breakdown of Democracy in Turkey</u>. Unpublished Ph.D. Dissertation. University of Michigan at Ann Arbor.

7

The Food Crisis in Kenya

Taye Woldesmiate and Ron Cox

INTRODUCTION

An overview of the Kenyan peasantry indicates that the majority of farmers face a marginal existence which appears increasingly problematic in light of recent economic trends. Kenya's Gross Domestic Product (GDP) has decreased from an annual growth rate of 4.6 percent in 1964 to below 3 percent by early 1980s. Kenyan foreign reserves continue to fall and are good for only two months of imports. Meanwhile, from 1974 to early 1980s foreign development assistance averaged 250 million dollars per year, with the annual disbursement increasing by 20 percent each year. The development assistance has coincided with increasingly tough IMF conditionality urging an increased commitment to private enterprise, foreign investment, free trade and lower government spending.

This paper examines some of the most important recent trends in Kenyan food production. Our central theme is that IMF and government policies have exacerbated a food crisis in Kenya characterized by a steady decline in stable food production over the last decade. We begin our analysis with a brief historical overview of British colonialism in an attempt to locate the current food crisis in proper perspective. Then, we attempt to show how the majority of peasant farmers are unable to rely on subsistence farms to feed their families. Much of the Kenyan peasantry are forced to depend on two or more jobs to generate the income needed to buy food on the free market. The population shifts in Kenya can be explained in part by the efforts of peasants to earn the amount of income necessary to survive. Some of the best land is now being used for the production of export crops including tea and coffee. The profits from production of export crops are doing little to alleviate the increasing unemployment, malnutrition and hunger in Kenya. Landless peasant farmers struggle to find work which will pay enough to buy food that was once grown at home and is now imported at an increasing rate. Other peasants must divide their time by working on both their subsistence farms and on the plantations owned by large corporations and wealthy Kenyan elites. At the same time labor power is needed to manage the subsistence farms, Kenyan males must look for industrial work in Nairobi or agricultural jobs on private estates. The result has been a decline in food production that threatens to further increase malnutrition and hunger in Kenya.

HISTORICAL BACKGROUND

Kenyan food production can best be understood by locating the current food crisis in proper historical setting. Kenya was originally held under a charter company, the Imperial British East Africa Company, which enjoyed a wide range of power that included the ability to recruit its own army. In 1893, this arrangement was replaced by European settlements. Africans, who before were purely subsistence farmers, were now forced to become laborers because of the monopoly of the European cash crop settlers. African production of profitable coffee, sisal and tea were at first discouraged and the made illegal (Amin, 1979).

The presence of white settlers in the colonial era had created an economy which exploited the agrarian products of coffee, tea, pyrethrum, and corn. After the seizing of 8.5 million acres of the most fertile land by military conquest, the British colonial apparatus put the expropriated African population onto reserves with loyal Africans appointed as chiefs. The racially hierarchical structure of the colonial society was reinforced by the importation of Indian traders who distributed the commodities of the British corporations while others serviced the middle class (Amin, 1979).

By 1924, six million acres or twenty percent of Kenya's best land were expropriated for European settlements and were primarily held by two syndicated and five wealthy individuals. At this time, 1,715 settler farmers employed 87,000 African laborers (Dumont, 1976). During the 1920s and 1930s, 2,000-3,000 more Europeans settled on more Kenyan land. Plantation agriculture in Kenya expanded as more companies sought to control the total production process by not only manufacturing, marketing and distributing a product but also by growing the crop. In the early 1920s, Brook Bond, Ltd. started the first plantations of tea in Kenya. The corporation, with the aid of the colonial government, uprooted the indigenous population by force and drove them onto the poorer farming land. Unable to sustain themselves, the uprooted peasantry were forced to look for work in urban areas or on the plantations (U.N. Center, 1980).

Through the use of military force and legal decree, the British colonial state established a two-tier economic system whose legacy persists to this day. The first tier of this system includes the large capitalist farms which produce for the export market and control a disproportionate share of arable land. The second tier consists of the majority of African peasants dependent on small holdings for survival and forced to supplement their meager incomes with industrial and plantation wage-labor. The British colonial state encouraged the expansion of export agriculture and the displacement of peasant farmers through the imposition of a variety of coercive policies:

1. The initial military conquest of Kenya and the takeover of the White Highlands for exploitation by the European settlers.

2. The resettlement of Kenyan peasantry onto less arable land, thereby making survival imperative upon both maintaining the subsistence farms and working for the European capitalists.

3. The allocation of nearly one million pounds of aid to white settlers through the agencies of the colonial government.

4. The legal and social discrimination that prevented African farmers from growing profitable crops for the export market.

One scholar provides a cogent summary of European legal and social discrimination that came out of the above policies:

> There was persistent discrimination against Africans who tried to grow cash crops. That African farmers were prevented from planting coffee is well known, but the same was true of pyrethrum. All producers had to obtain a license to plant these two crops and the government simply refused to grant licenses to African growers. There was considerable political controversy over the refusal of coffee licenses and, therefore, in 1934 licenses were offered to a very small number of growers in the Meru and Kisii areas. This had little effect as Africans were afraid that

the Europeans would alienate their land if they planted coffee (Van Zwanenberg, 1975).

By the 1940s, the British had strengthened their legal decrees to prohibit African production of profitable sisal and tea. Meanwhile, the socioeconomic tensions between the British minority and the African majority reached a crisis point as a result of several factors. First, there was increased population pressure on the land as Africans were systematically displaced by larger European settlements. The colonial state concentrated on the modernization of European farms in the 1940s in response to increased opportunities to buy and sell in the world market. At the same time, the administrators continued to deny Africans the legal right to compete with European farmers in the production of the most profitable export crop. Second, the colonial state wanted to achieve two goals which only served to exacerbate the tension between the poor African peasant and the colonial administrator. Accordingly, the state sought to preserve the position of the capitalist European farmer by maintaining a reserve policy that discouraged Africans from producing export crops for profit. This, in turn, forced the majority of Kenyan farmers to supply the needed wage laborers for the newly modernized European capitalist tier of the economy. As a result, the Europeans benefitted from the maintenance of African subsistence farms. The state also sought to ease the population pressure on the land by developing a government program designed to prevent further deterioration of the soil through various conservation projects and the preservation of water supplies (Van Zwanenberg, 1975). In attempting to preserve the worst inequalities in the ownership of land, the British hoped to forestall rebellion by maintaining subsistence agriculture for the majority of Kenyans and by maintaining the profitable estates of the Europeans relatively free from competition with African producers.

Despite the legal restrictions imposed by the British, some Kenyans were able to become wealthy as a result of their access to limited amounts of land and employment during the colonial period. Kenyans were not allowed to compete with settler farmers by growing coffee. However, a few Kenyans with substantial landholdings could grow other cash crops like wattle or tobacco. Other Kenyans became shopkeepers, skilled workers, government clerks and teachers thereby earning higher incomes than most of their fellow countrymen. Some Kenyans enriched themselves by cooperating with British colonialists to help facilitate settlement schemes. A few received relatively large incomes as educated missionary converts who worked as teachers or clerics for the church. These individuals as a whole constituted an educated, upwardly mobile minority who became increasingly frustrated over the British legal and social decrees that prevented them from acquiring more wealth.

The socioeconomic tensions between the European administrator, the rising minority of Kenyan elite, and the majority of poor Kenyan small farmers resulted by the 1950s in a gradual formation of a nationalist movement to reclaim land from settlers and to rid Kenya of European political control. Increasing nationalist pressures prompted the government to abandon its reserves policy by 1952 and to permit wealthy Kenyans to purchase land for profit in areas that had previously been the preserve of the Europeans. From 1952 to Independence in 1963 there was an orderly transition of large-scale farms from European ownership to wealthy, indigenous Kenyans who soon had a vested interest in maintaining large estates. After independence, these wealthy Kenyans (consisting mainly of a small percentage of the Kikuyu tribe) gained key positions of political influence in the Kenyan state. Their stated aim was the continued promotion of large-scale farming through direct monetary assistance to plantation farmers and tax breaks to multinational corporations. In other words, independence did little to alter the concentration of land ownership or to change the direction of government policy, which remained committed to the interests of corporate farms, despite the fact the vast majority of Kenyans were dependent on small farms for their existence. As Van Zwanenberg (1975) note:

> Five percent of White Highlands was taken over by small-scale farmers, but most of the coffee, tea and sisal plantations and the cattle ranches remained intact. What had changed was the nationalities of the owners. Wealthy, indigenous Kenyans, including several men well-known in public life, bought farms from the Europeans. The Land Bank, which advanced loans in the 1930s to Europeans, now lent money to the new farmers...It is interesting to note that it was the areas of highest productivity in the former White Highlands which were not touched by the resettlement schemes. The core of large-scale, mixed farming, the basis of the conflict over economic resources, has been retained.

As we will see in the following section, the legacy of the colonial agricultural policies continues to this day as a result of government policies that continue to favor the largescale Kenyan farmers at the expense of the smallfarmer. These policies remain intact despite the fact that 77 percent of all Kenyans depend on access to smallholdings to feed their families. The erosion of these smallholdings has meant a continued decline in subsistence food production as more land is bought by large companies and used for export agriculture.

THE COLLAPSE OF THE FOOD SYSTEM

The collapse of the Kenyan food system is related to several factors. First, land in Kenya is highly unequally distributed. It has been estimated that 0.1 percent of holdings contain 14 percent of the arable land, while 2.4 percent of the holdings contain 32 percent (Hunt, 1984). While 30 percent of smallholdings are under half a hectare, and 54 percent under one hectare, the largest farms exceed 1,000 hectares. Basically, there are three main categories of farms in Kenya: the smallholding sector which occupies about 3.5 million hectares in the late 1970s between about 40,000 holdings, the "gap" farms which occupy about one million hectares divided between about 40,000 holdings, and the large farms which occupy about 2.1 million hectares: of these 2.1 million some 0.53 million hectares are in mixed grain and dairy farms, about 1,200 in number, and 1.64 million are in plantations and ranches (Livingston, 1981).

Second, the Kenyan government has encouraged this unequal distribution by providing credit and resource allocation to the large farmer and virtually nothing to the Kenyan smallfarmer (Race and Class, 1983). The stated rationale is that large farmers are more productive and will maximize profits in a way that encourages savings and reinvestment. In reality, the large farmers, with access to government credit, resources and technology, have been among the least productive in the utilization of crop land. For example, half the land belonging to large farmers is lying idle. According to our findings, the reason for this lack of productivity is that large farmers choose to invest their profits and savings in areas other than agriculture. Many choose to engage in speculation or investment schemes that are unproductive. Meanwhile, the large landowners simply maintain the idle land as a status symbol in the short run and as a potential salable entity in the long run. As a result, land that could be used to grow food crops for domestic consumption is left idle while the vast majority of Kenyan farmers struggle to survive on the margins of the least arable land.

Third, foreign-owned companies continue to exercise substantial control over the production, marketing and distribution of agricultural produce in Kenya. The largest twenty companies have an 86 percent share of all the national business investments in Kenya. More than 60 percent hold a virtual monopoly in the main product they sell (Langdon, 1978). This is the case despite the fact that foreign companies produce commodities for export that cannot be used to feed the Kenyan population. Race and Class (1983) documented the effects of corporate investment and the links between corporations and government policy:

> Foreign companies still control thousands of acres of tea, coffee, sugar, sisal, fruit plantations and ranches. The government appears more than willing to continue putting the agricultural resources of Kenya at the service of foreign capital. For instance, it obligingly killed smallholder production of pineapples around Thika in order that the American [Del Monte] Company could have a monopoly. Later it negotiated a new deal with Del Monte, promising to exempt the company from whatever change might occur in foreign investment policy during a twenty-five year period.

Thus, government policy works to support the interests of export agriculture and the large farmer despite the fact that majority of Kenyans (about 88 percent of the population) are either landless or dependent on smallholdings for survival. A percentage breakdown of the Kenyan farming population illustrates the precarious nature of their existence and helps shed light on some of the demographic trends leading to the reduced production of food. Twenty-two percent of the Kenyan rural population are landless; 44 percent are smallholders with less than seven acres on which to grow crops for their own use and for sale; and the remaining third are potentially better off with seven or more acres of land. The rural landless struggle to find temporary employment on plantations or in the manufacturing sector at the same time the economy faces a slowdown of activity. For example, agricultural wage employment fell from 37 percent to 23 percent between 1964 and 1980 (Bureau of Statistics, 1981). The share of the landless have begun to migrate from rural to urban areas in search for new employment options.

Unable to find agricultural work, the rural landless increasingly turn to the manufacturing sector for employment. Yet, manufacturing has undergone a slow expansion since 1963 from 13.3 percent of GDP to 14 percent in 1980 (Bureau of Statistics, 1981). If this kind of growth rate continues, the landless poor cannot hope to find the employment necessary for survival. As O'Keefe (1984) notes, Kenya's potential labor force is projected to rise from 6.1 million to 15.7 million by the end of the century. At the same time, existing data indicate that nearly one-third of rural poor falls below the poverty line "based on expenditures required to maintain an adequate nutritional level" (O'Keefe, 1984).

Faced with a lack of employment opportunities in the agriculture and manufacturing sectors, the rural landless are burdened by severe constraints in attempts to alleviate their poverty. Some have become streetwalkers, shoeshine boys, smalltime mechanics and producers of cheap furniture and shoes in efforts to earn a livable wage. The fact that these jobs are inherently unstable and pay little contribute to the legitimate concern that the fate of the rural landless is increasingly precarious. In addition, the prospects of the landless acquiring land for subsistence or export production appear equally futile. As O'Keefe (1984: 160) notes:

> During the late 1960s and early 1970s, population growth was accommodated by heavy rural to rural migration towards the more marginal areas. The spacial limits of that migration have been reached since it is impossible to farm areas where the probability of drought exceeds 40 percent. In the more densely populated areas, available land per capita will fall from 0.35 of a hectare in 1979 to 0.24 of a hectare in 1989. By the end of the decade, only 0.05 hectares per person of medium potential land will be available.

The second category of the rural population are those farmers who own less than seven acres of land. These smallholders often depend on off-farm sources for more than half their income. As a result, they are dependent on job opportunities on plantations or in the urban manufacturing sector to supplement their agricultural production. It is important to keep in mind that this group constitutes 44 percent of the Kenyan peasantry. The ability of these smallholders to maintain their agricultural production helps determine whether or not these smallholders rely on subsistence production to provide some of their food needs. This production is increasingly threatened by the labor demands of the market economy. At the very time that labor is needed to maintain the smallholder farm, the male member of the household is working on large plantations for foreign-owned companies or Kenyan elites, or is working in the manufacturing sector in the urban areas. This absence of the male members of the households, at a time when the smallholder farm needs attention, has been one of the factors causing a consistent decline in food crop production over the last decade as shown in Table 7.1.

In addition, there is an incentive among smallholder families to have more children in order to increase the labor power available to manage the small farms. This is due to the fact that the male is often absent from the land at the very time terrace construction is needed to prevent soil erosion. The decision to have more children means more potential laborers to manage the duties of the farm. Yet it paradoxically also means more overcrowding and more mouths to feed. The smallholder is thus faced with a virtual no-win situation. To compound the problem, the smallholder has little access to extension services or government inputs like fertilizers and seed, which might help raise productivity. In the midst of persistent labor pressures and lacking governmental assistance, these farmers are faced with an increasing erosion of their soil and a dramatic reduction in the amount of land used to grow food crops.

If the rural population were earning enough income from plantation wages or the manufacturing sector, then they could possibly afford to buy the food necessary to feed their families. Indications are, however, that there is a growing disparity between the incomes of the rural and urban populations of Kenya. The urban sector, with only 10 percent of the population, earns 43 percent of the income. The rural sector, with 90 percent of the population, earn 57 percent of the national income. One-third of the rural population falls below the poverty line and the rural areas contain 98 percent of the poor (Livingstone, 1981). In other words, the temporary plantation and manufacturing work available to rural families does not provide them with enough income to rise above the poverty line.

The serious impact of rural poverty on the Kenyan economy is illustrated by the rural to urban migration of the peasants. As more land is used for plantation

Table 7.1
Agriculture and Food Production in Kenya, 1973-1984

Commodity (1,000 metric tons)	1973	1974	1975	1976	1977	1978	1979	1980	1981	1982	1983	1984
wheat	150	149	158	176	178	166	201	185	210	225	205	80
rice, paddy	36	33	32	40	43	42	37	40	41	38	36	38
corn	1,600	1,600	1,900	2,195	2,205	1,895	1,450	1,750	2,200	2,340	2,070	1,700
barley	34	31	34	32	34	48	60	75	82	100	100	90
oats	6	5	5	8	9	9	7	7	7	7	8	6
millet	320	360	375	351	350	351	296	350	350	197	300	195
beans, dry	34	29	41	16	25	20	25	23	25	26	27	20
potatoes	304	320	336	342	341	361	360	350	365	346	260	235
cotton	5	5	5	5	5	9	9	13	14	9	8	7
cottonseed	10	10	10	10	11	18	18	26	28	18	16	14
peanuts, in shell	1	2	3	6	8	8	8	8	8	8	9	7
sesame seed	2	1	3	4	7	7	8	8	8	9	9	8
sunflower seed	2	4	5	9	13	14	15	15	15	16	17	17
castor beans	1	1	2	2	2	2	3	3	3	3	3	2
pyrethrum flowers, dried	11	14	15	14	11	8	8	11	18	19	9	7
vegetables	384	404	308	368	394	409	421	427	431	440	450	410
cashews, in shell	15	10	11	23	15	10	11	16	10	12	8	12
pineapples	48	44	83	100	125	130	143	145	150	155	160	145
coffee	71	70	66	75	101	85	74	92	100	89	92	126
tea	57	53	57	62	86	93	99	90	91	96	120	116
sisal	58	86	44	34	33	31	37	47	41	50	51	38
sugar, raw	171	176	172	161	194	255	314	421	287	325	350	375
meats	180	182	184	208	246	261	285	288	308	303	316	326
milk	900	850	927	898	934	946	860	810	850	1,150	1,300	1,150
hides and skins	25	24	25	28	35	37	40	42	43	44	46	47

Source: USDA, World Indices of Agricultural and Food Production, 1973-1985.

agriculture, still more Kenyan peasants will be forced to move in growing numbers to the city in an effort to supplement their meager incomes. It is estimated that the Kenyan labor force will rise from 6.1 million to 15.7 million by the turn of the century. Meanwhile, wage employment has slowed down from a growth rate of 3.9 percent between 1964-1974 to 3.3 percent between 1974-1980. Even if the labor force rises by six percent per year, an estimated 12 million people will still be unemployed by the turn of the century (Statistical Abstract, 1980).

It is important to note that the category of the rural population with seven or more acres of land are slightly better off than peasants with less or no property. Yet, even this group suffers from a marketing and distribution scheme that favors middlemen at the expense of producers in the setting of prices for agricultural products. Numerous Kenyan cooperatives have been established since Independence to control the production of agricultural crops by setting artificial quotas and by regulating the prices of crops. These cooperatives are managed by Kenyan governmental officials who make the decisions regarding production levels and price for peasant-produced crops. Cooperatives in the 1970s marketed more than half of all coffee, 40 percent of all milk and the entire pyrethrum output (<u>Race and Class</u>, 1983). Kenyan government officials pay peasant producers an initial price for their produce, store the crop temporarily in government warehouses, and set another substantially higher price for the crop when it is sold on the market. In this way government officials are often able to make a significant profit via the difference between the market price and the price paid to peasant producers. The final price is set high enough to pay the middlemen or the Kenyan bourgeoisie who markets the product. Thus, this artificial pricing allows the marketers and distributors of agriculture produce to reap the bulk of the profits from the peasants.

This marketing scheme has negative implications for the distribution of food crops in Kenya. For example, a recent drought in 1981-1983 saw a subsequent decline in the production of corn and other staple food. A famine occurred partly as a result of the fact that, prior to the drought, the National Cereals and Produce Board had sold the bulk of surplus corn to foreign buyers. During the late 1970s, the Board bought corn at cheap prices from the Kenyan peasantry. In 1978 and 1979, the Board exported nearly 20,000 tons of corn to foreign buyers in order to secure profits for the Board's directors. The export sale diminished virtually the entire Kenyan surplus of corn despite the fact that domestic demand for food was high. In addition, the corn surplus was directly needed during the drought of the early 1980s as a partial replacement for the dramatically reduced production of corn.

The low prices for crops have also discouraged the production of food crops by the Kenyan peasantry. Realizing that they will be paid little for their efforts, peasants have dropped out of cooperatives in increasing numbers. For example, membership in the Mathira Dairy Cooperative dropped from 11,254 to 2,000 between 1981-1983. In an atmosphere of reduced incentives for the small farmer to produce food commodities, food security in Kenya would continue to be threatened in years to come.

FOOD vs CASH CROP PRODUCTION

Kenyan state support for multinational corporations and the Kenyan bourgeoisie has had a direct impact of the agricultural sector of the economy. The development strategy of the Kenyan government has favored export-led growth, with emphasis on cash crops over the production of food crops by smallholders. As outlined earlier, Kenyan smallholders face a precarious existence due to their reliance on two or more jobs to survive in the market economy. As a result, many smallholders have been forced by the necessities of the market to neglect farms that once produced food crops. Many abandon food crops entirely and contract themselves to MNCs or state cooperatives in the production of cash crops. The result has been a growth in export crop production and a dramatic decline in Kenyan food production.

The tradeoff between food and cash crop production is shown in Figure 7.1. The data demonstrate that production of major export commodities of pineapples, coffee, tea, and sugar have dramatically increased between 1973 and 1984: 202 percent increase in pineapple products, 77 percent increase in coffee, 103.5 percent increase in tea, and 119 percent increase in sugar production. It is crucial to note that, for the same time period, food crops (wheat, rice, beans and corn) do not show similar trend. Instead, there is sharp decline in corn and wheat production in recent years. Rice production remained about constant during 1973-1984, increasing only minimally from an index value of 100 in 1973 to 101 in 1984. The decline in bean production began in 1976 and continued to remain below the 1973 figure.

One of the reasons for the production declines during 1982-1983 is a severe drought that damaged crops and

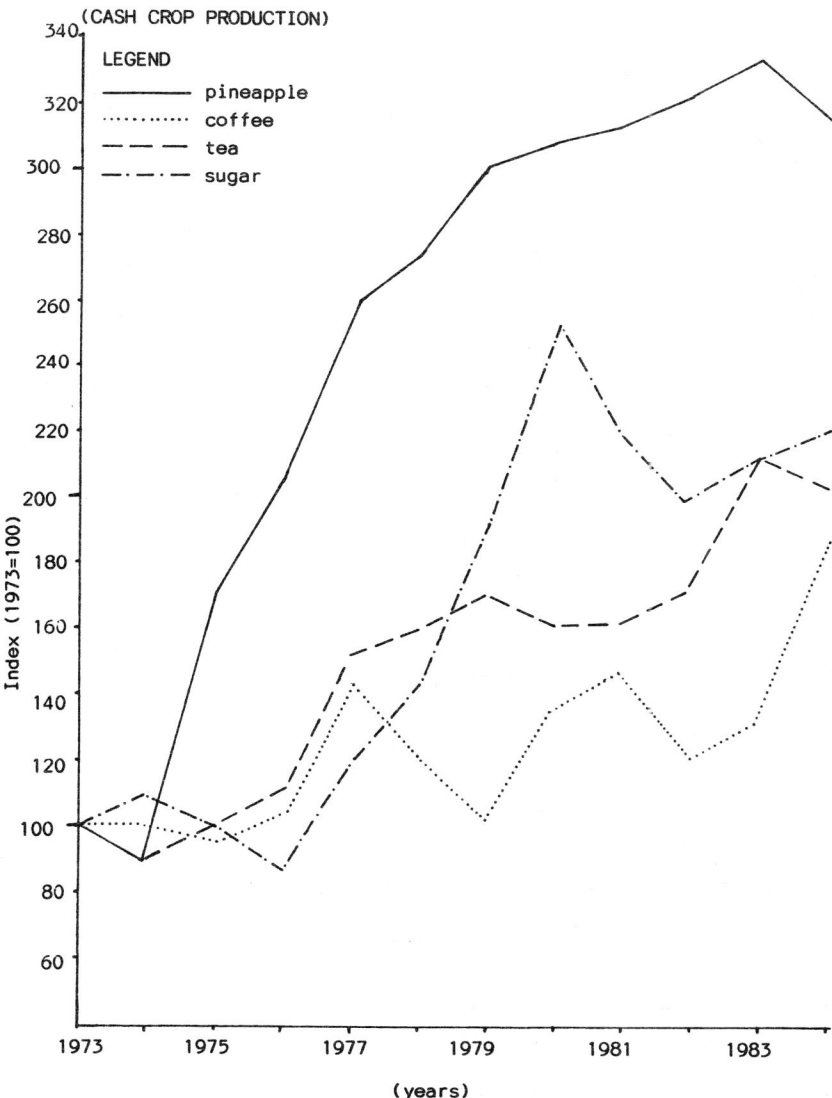

Figure 7.1 Index of Agricultural Production in Kenya

(continued)

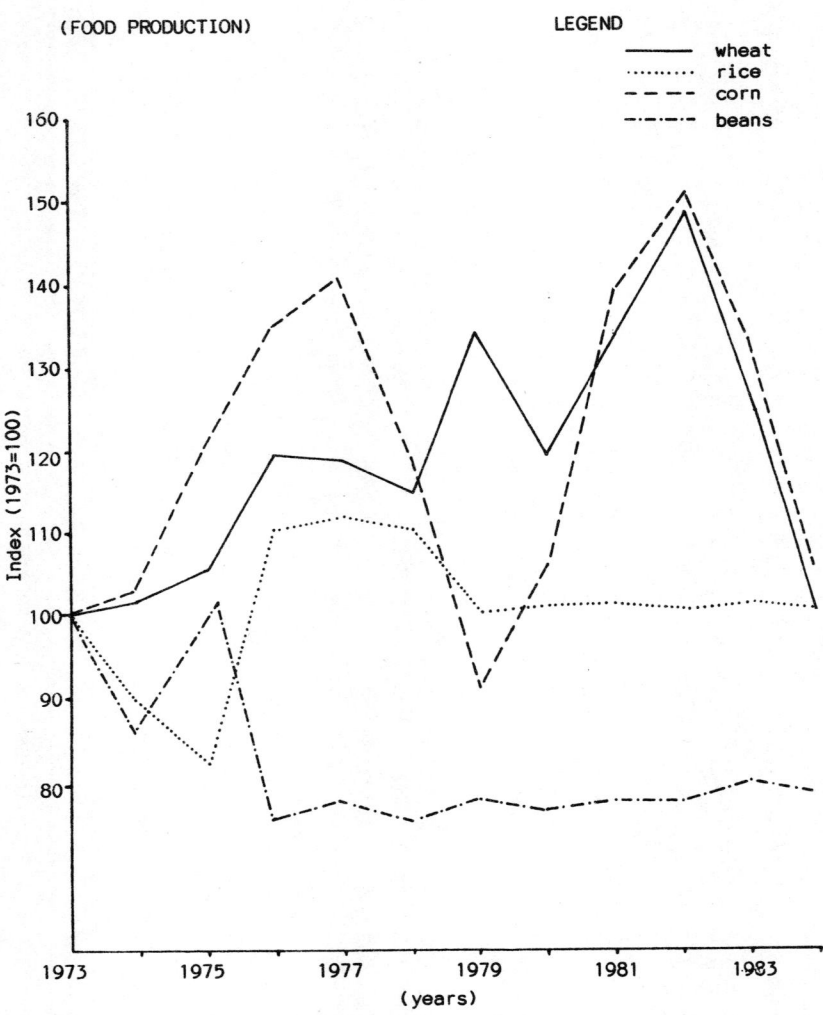

Figure 7.1 Index of Agricultural Production in Kenya

Source: USDA. World Indices of Agricultural and Food Production.

contributed to famine in some parts of Kenya. Yet, the drought, by itself, does not explain the continued decline in production since 1983, nor the reason for the relative success of the export crop sector. The food crops were most severely damaged by the drought since their cultivation was largely performed by peasants who had little or no access to government inputs such as irrigation systems. In addition, as already discussed, many peasants by the 1980s, had begun to leave food cooperatives as a result of a pricing scheme which paid them little for their labor. Peasants that did remain on their land were faced with severe soil erosion that significantly impaired production (O'Keefe, 1984). On the other hand, large farmers specializing in key export crops were able to withstand the worst effects of the drought as a result of their access to adequate irrigation systems and credit to help insure cultivation and production of cash crops.

The rationale for the promotion of export sector of the economy is a development strategy which urges a maximization of exports in an effort to generate significant foreign exchange needed to promote economic growth and development. This strategy has been followed by Kenyan leaders since independence and has been reinforced by the monetary assistance from the World Bank and IMF, which also recommend such a strategy. This strategy, as will be discussed in the next section, depends on the notion that the most productive sectors of the economy will invest and reinvest, thereby allowing for a multiplier effect that will create jobs and spur growth. However, despite the Kenyan government's efforts to promote the production of cash crops, the trade deficit has remained high since 1977 as indicated by the data in Table 7.2.

The dramatic drop in export earning in the first part of 1980s indicates the degree to which the Kenyan economy is subject to the vagaries of the world market conditions. The world recession of 1980-1983 meant that Kenya faced a significant shortfall of foreign exchange. Couple this with the drought of 1982 and you have a country that must borrow significant amounts of money to ease its balance of payments crisis. In order to alleviate this problem the Kenyan government signed a series of agreements with the IMF and borrowed extensively from the international financial agencies. The result, however, was an additional burden that was to create more problems for the economy. This new burden was the growing debt service ratio which increased from 1.1 percent in 1975 to 10.5 percent in 1979 and continued to increase reaching 21.5 percent in 19083. This has

Table 7.2 Balance of Payments and External Debt for Kenya
(at current $U.S. million)

	1973	1974	1975	1976	1977	1979	1980	1981	1982	1983	1984
Balance of payments											
exports	679	993	952	1,095	1,556	1,571	1,959	1,659	979	876	1,028
imports	733	1,171	1,135	1,092	1,436	1,972	2,837	2,295	1,683	1,274	1,547
balance	-54	-238	-183	+3	+120	-401	-878	-637	-704	-398	-469
External debt	479	614	1,127	1,396	1,863	3,035	3,038	3,404	3,675	3,680	3,691
Debts servie as percent of exports	1.5	2.1	1.1	5.8	5.1	10.5	12.7	17.1	22.1	21.5	27.9

Sources: World Bank, World Tables (1980, 1983), World Development Report (1985).

meant that the debt/GNP ratio for Kenya has increased from 25 percent in 1970-1972 to 43 percent in 1980-1982 (World Bank, 1985: Figure 4.1). For a developing country like Kenya, such debt figures are quite alarming as they indicate that unless the trend is reversed soon, the export earnings will only meet the economy's external debt obligations. Under these conditions the government would continue to find it extremely difficult to allocate resources for essential domestic development needs.

THE ROLE OF THE IMF

Despite the economic problems it faces, Kenya has received high marks from the IMF in the administration of its foreign assistance program. In Kenya, and virtually everywhere else, the IMF follows a liberal trade and monetary ideology designed primarily to liberalize exchange and import controls and to facilitate in introduction of foreign capital in the domestic production process. These policies emerge from the monetarist school of thought that guides the IMF. Following the adoption of the IMF policy package by a given government, that particular state receives credits from the Fund and a "green light" is signaled to other international financial institutions to extend similar credits, if needed, to the country concerned. Kenya signed three such agreements with the IMF, latest being in 1984, and as a result additional credits accumulated a 3.1 billion dollar debt. The IMF austerity packages adopted by the Kenyan government in 1978, 1981, and 1984 emphasized the following policy suggestions:

1. Limitations on the level of domestic spending, often entailing wage ceilings, reductions in employment, and elimination of state subsidies on consumer goods.
2. Expansion of government revenues through additional or more effective tax collection.
3. Elimination of inefficient or unprofitable public enterprises.
4. Ceilings on credits, domestic and external borrowing and monetary expansion.
5. Redirection of credit away from the public to private sector.
6. Devaluation of the national currency.
7. Relaxation of price controls.
8. Removal of exchange and import controls in all sectors of the economy (Browne, 1984).

The program aid offered by the IMF has stressed the above measures as a combination of policies and goals to be followed and attained by the Kenyan government. The policy-stricken bear the burden of these austerity programs with significant reductions in already meager incomes, further widening the gap between rich and poor. The goal of the IMF is tied to freeing the export economy from governmental restraints to provide more income for necessary imports. As a result of these policies the national bourgeoisie and the debtor state became more dependent on external market conditions.

THE RISE OF MNC: A CASE STUDY OF DEL MONTE

Pineapples are becoming a major Kenyan export crop and plantations are producing more each day. The government encourages the industry in order to reduce its reliance on tea and coffee as means of acquiring foreign currency. Del Monte, which is now owned by the R.J. Reynolds food and tobacco conglomerate, is the major MNC in Kenya that produces pineapples today.

Before Del Monte became involved in the pineapple trade, two businessmen established a pineapple firm in Kenya in 1949. Fresh fruit was supplied to Nairobi through this firm, but in 1958 it was taken over by the Tancol group. This firm was originally developed as a small farmer scheme. Because of its growth, more small farmers were attracted to it. However, complaints arose concerning a limit restricting the growth of pineapple to approximately a 25-mile radius around the processing plant. The desire for pineapple grew and farmers responded quickly, but within three years, the company was complaining that they were receiving more pineapples than they could handle. Thus, the government began to seek a MNC with management experience to handle the industry (Kaplinsky, 1979).

The Kenyan government invited Del Monte, a U.S. based MNC, to take over the pineapple industry. Traditionally, Del Monte grew pineapples in Hawaii but the workforce there had become increasingly more expensive as they had negotiated large salary raises in early 1960s. This and raising land prices in Hawaii prompted the company to expand production in the Philippines and to look for alternative sources of supply. One of these alternative sources was Kenya. The government and the people of Kenya were in need of jobs, tax revenues, and much desired foreign exchange.

The company signed a 33 year contract with 95 percent ownership rights with the Kenyan government in 1965.

This takeover of the pineapple industry in Kenya by Del Monte produced 4,500 jobs and brought in 20-30 million U.S. dollars annually in export revenues. However, the tax revenues created by this arrangement has not been profitable. According to Kaplinsky (1979), the company rigged export prices and cooked its accounts in order to prevent taxation by the Kenyan government. The scheme was a classic example of transfer pricing. Del Monte's Kenya canners sold its pineapple to Del Monte's British subsidiary on paper at ridiculously low prices. The British company then resold the goods in Europe at the market price. Furthermore, in order to create complete Del Monte monopoly in pineapple sector, Del Monte persuaded the Kenyan government in 1965 to issue a regulation restricting planting of pineapples to those areas originally licensed by the Canning Crops Board. Through this action, the participation of smallholders in pineapple production was greatly reduced.

Under the 1965 agreement by which Del Monte managed Kenya Canners, the government agreed to purchase an old sisal estate at Thika and lease it in sections to the company at a lower annual rate. In return, it was understood that Del Monte would develop their own estate as well as take supplies from the surrounding smallholders. In 1968, the year when Del Monte actually bought Kenya Canners, they came to the conclusion that smallholder production would have to be discouraged and pineapple would then be grown entirely on estates. Del Monte argued that estate production suited their international standards of production better than smallholder production. This process would allow them to control the quality and size of the pineapple and it also suited their management needs. In face of these developments the Kenyan government chose to support the company's position due its desire to increase export earning of Kenya. However, the Kenyan government did not receive the revenues it had hoped for from Del Monte sales. The modern production procedures used by the company required imports of heavy machinery, fertilizers, pesticides, and herbicides. Even sugar, which is used in the canning process, was imported. The main rationale behind sugar imports was the low quality of domestic sugar which made it unsuitable for Del Monte's needs. The cans were produced locally but tin was imported. This process is still in effect in Kenya. In addition, foreign companies have the right to remit profits under the Foreign Investment Act and Del Monte's U.S. parent company had managed to obtain a technological agreement which gives it a commission of five percent of the net sales.

CONCLUSION

This brief overview has attempted to highlight the growing burdens faced by the Kenyan peasantry in the midst of a shift from subsistence to commercial agriculture. As recent studies have shown, the rural poor are migrating from their homes to urban areas in search of wage employment to supplement a declining subsistence base. Meanwhile, large landowners have increased their holdings of plantation farms producing export crops for the world market. Many peasant families are forced to take two or more jobs to earn the income necessary to feed themselves. Even so, most have not risen above the poverty line and are poorer now than they were two decades ago. The IMF austerity measures have resulted, among other things, in increased prices for food. In addition, the government cooperatives continue to raise prices to pay the salaries of middlemen. This has further discouraged production of food crops due to the low prices peasants can expect to receive for their produce. At the same time, the government has encouraged corporations like Del Monte to virtually eliminate smallholder production of pineapples in favor of estate farms that bring little less than the desired amount of revenue into the country. This has further increased the deficit, making it still more difficult for Kenya to purchase the imports it desperately needs to feed its population.

The IMF austerity measures as adopted by the Kenyan government have promoted the interests of international capitalism through external loans, investments by MNCs and the production of cash crops, all of which inhibit the development of Kenya's domestic food production. The outcome has been increasing dependency on foreign exchange to purchase the vital imports which no longer be supplied locally. The Kenyan government has worked with the IMF to secure a climate of security for foreign investment on the backs of the poverty stricken peasants. As the partial proletarianization of the peasantry intensifies, the state machinery will be faced with an increasingly landless population unable to maintain themselves and demanding fundamental changes in Kenyan politics.

REFERENCES

Amin, Samin (1979). <u>Colonialism in Afric</u>.
   New York: Monthly Review Press.

Bureau of Statistics (1981). <u>Statistical Bulletin</u>.
   Nairobi: Kenyan Government Publications.

Browne, Robert S. (1984). "Africa and the IMF: Conditionally a New Form of Colonialism." <u>Africa Report</u>. (September/October).

Dumont, Rene (1976). <u>False Start in Africa</u>.
   New York: Preager Publishers.

Hunt, Diana (1984). <u>The Impending Crisis in Kenya: The Case for Land Reform</u>. Vermont: Grower Publishing Co.

Kaplinsky, Raphael (1979). <u>Multinational Corporations in Kenya</u>. Oxford: Oxford University Press.

Kenyan Cental Bureau of Statistics (1985). <u>Economic Statistics</u>. Nairobi: Central Bureau of Statistics.

Langdon, Stephan (1978). <u>Multinational Corporations in the Political Economy of Kenya</u>. Oxford: Oxford University Press.

Livingston, I. (1981). <u>Rural Development, Employment and Incomes in Kenya</u>. Addis Ababa: Artistic Printing Press.

O'Keefe, Phil (1984). "Poverty, Proletarisation and the Production of Uneven Development: A Kenyan Village." In B. Muslow and H. Finch. eds. <u>Proletarisation in the Third World</u>. London: Croom Helm.

U.S. Department of Agriculture (1983). <u>World Indices of Agricultural and Food Production, 1973-1982</u>. Washington D.C.: USDA.

_____ (1984). <u>World Indices of Agricultural and Food Production, 1984-1983</u>. Washington, D.C.: USDA.

U.N. Center of Transnational Corporations (1980). <u>Transnational Corporations in Food and Beverage Processing</u>. New York: United Nations Publications.

Van Zwanenberg, R.M.A. (1975). An Economic History of Kenya and Uganda, 1800-1970. London: The MacMillan Press.

World Bank (1980). World Tables. 2nd ed. Baltimore: Johns Hopkins University Press.

_____ (1983). World Tables. 3rd ed. Baltimore: Johns Hopkins University Press

_____ (1985). World Development Report, 1985. New York: Oxford University Press.

PART 2

INTERNATIONAL FACTORS AND AGRICULTURE IN THE THIRD WORLD

8

The Impact of Dependency on Agriculture and Food Crises in the Third World

Birol A. Yeşilada

In the last two years the eyes of the international community have once again focused on agriculture and food crisis in the Third World as the plight of the African countries facing mass starvation came to the forefront. These problems, however, are not unique to the African continent. Most developing countries face varying levels of malnutrition and hunger. Domestic government policy failures as well as international factors have resulted in inhibited agricultural development in the Third World.

According to the dependency perspective, the growth of peripheral economies (Third World) depends largely upon steady expanding trade and capital flows, both of which are closely related to world marker activity. World market activity. World market activity, on the other hand, largely depends on growth/decline in the economies of the core states, that is, the industrial West. In this context, dependence explains the external reliance of a state on another in a specific issue area. Examples of such relations include the dependence of the peripheral countries on the continuations of financial flows from the West. Furthermore, as Caparaso (1978) suggests, dependency represents the historical process through which peripheral states are incorporated into the world capitalist system and the socioeconomic and political structural distortions resulting from this process in these countries.

In this sense "Agrarian Reform in Reverse" can be understood as an outcome of Third World countries' attempts to carry out domestic development plans as their economies become influenced by external forces. Such factors include the international debt burden of most Third World states, developed countries' mercantilist trade policies towards the South's primary commodity exports, and the increasing influence of the IMF in reformulating the development policies of their countries. Faced with such problems, developing countries opt for policies aimed at generating quick cash, such as, production of cash crops and investments by international agribusiness companies. The outcome of these developments is the weakening of such countries' domestic food base and increased poverty among the rural population.

The dependency relationship between the core and periphery has been outlined by the early dependentistas (Bagu, 1949 and 1960; Cardoso, 1962; Dos Santos, 1967; Frank, 1966; Prebisch, 1959; and Furtado, 1964a). However, it was not until the 1974 United National Conference on Trade and Development (UNCTAD) that developing countries presented a united front on the issue of dependency. The New International Economic Order (NIEO) adopted by the Group of 77 developing countries at this conference called for reforms in the existing trade, monetary, aid, and investment relations between the core and periphery (UNCTAD, 1974). One of the issues emphasized in the NIEO was the call for preferential treatment of the peripheral states' exports, including agricultural goods, to the industrial West. Despite the continuous North-South conferences on these matters since UNCTAD in 1974, it is questionable whether the peripheral states' position has been improved. In order to determine this, it is essential to examine the casual factors the effect agricultural development in the Third World.

Since previous sections of this book have dealt with individual case studies that outline various domestic constraints on agriculture and food production in the Third World, this chapter will examine the international factors that influence agrarian development in the South. These are: falling agricultural commodity prices, unfair trading policies, the international debt crisis, and the increasing influence of the International Monetary Fund (IMF) in shaping development policies of Third World countries.

THE PROBLEM OF FALLING COMMODITY PRICES

Continuous decline in nonoil primary commodity prices have caused one of the central debates between North and South. Earlier, Raul Prebisch (1959) outlined two key dimensions of the debate for the peripheral states. The first was fluctuations in world markets that destabilize the export prices of primary commodities. And the second dimension was the deteriorating terms of trade due to lower income elasticities for primary commodities as compared to manufactured or semi-manufactured products. As Prebisch outlines, these problems are strongly dependent upon the market power of the industrial countries since the North has always been the major consumer marker for the South's exports. Since Prebisch's work, the nature of falling commodity prices (in real terms) has not changed with the obvious exception of oil. As shown in Figure 8.1 the value of the composite index the 33 essential nonoil primary commodities1 of the Third World countries stood at 73.0 in 1985. This figure was the lowest value for these commodities since World War II. It is clear, then, in those Third World countries where a major portion of foreign exchange earnings depend upon exports of these commodities, declines in the export-value of their goods with the dramatic increases in the value of oil imports have resulted in serious balance of trade problems as early as 1974.

Thus far, no comprehensive study has been carried out about the relative weight of the different casual factors behind the decline of nonoil commodity prices. Some of the major factors cited are: (i) deepening recessions in the world market. While the 1974/75 recession was the deepest in scope since World War II, the 1980-1983 recession was the longest in duration affecting almost all countries (Yesilada, 1987); (ii) the recent recoveries' bias in favor of services and high technology industries that do not consume much raw materials; and (iii) increasing costs of maintaining stocks of agricultural products at times of low market demand (World Bank, 1986). As Abramovic (1986: 956-957) explains, another general factor presses on the commodities:

> ...[this is] transfer losses, in the form of falling export prices, caused by an inelastic need for foreign exchange resulting from debt service committments and critical import requirements. Export sales are maintained in the face of heavy competition, insufficient foreign demand, and frequent trade obstacles in order to acquire desperately needed foreign exchange. A large number of primary producing countries are devaluating one after another in order to be able to compete at the present low world prices, but, as demand is frequently depressed and does not respond to price cuts, the main result of devaluations in these cases is to reduce real wages all around and further to depress world prices.

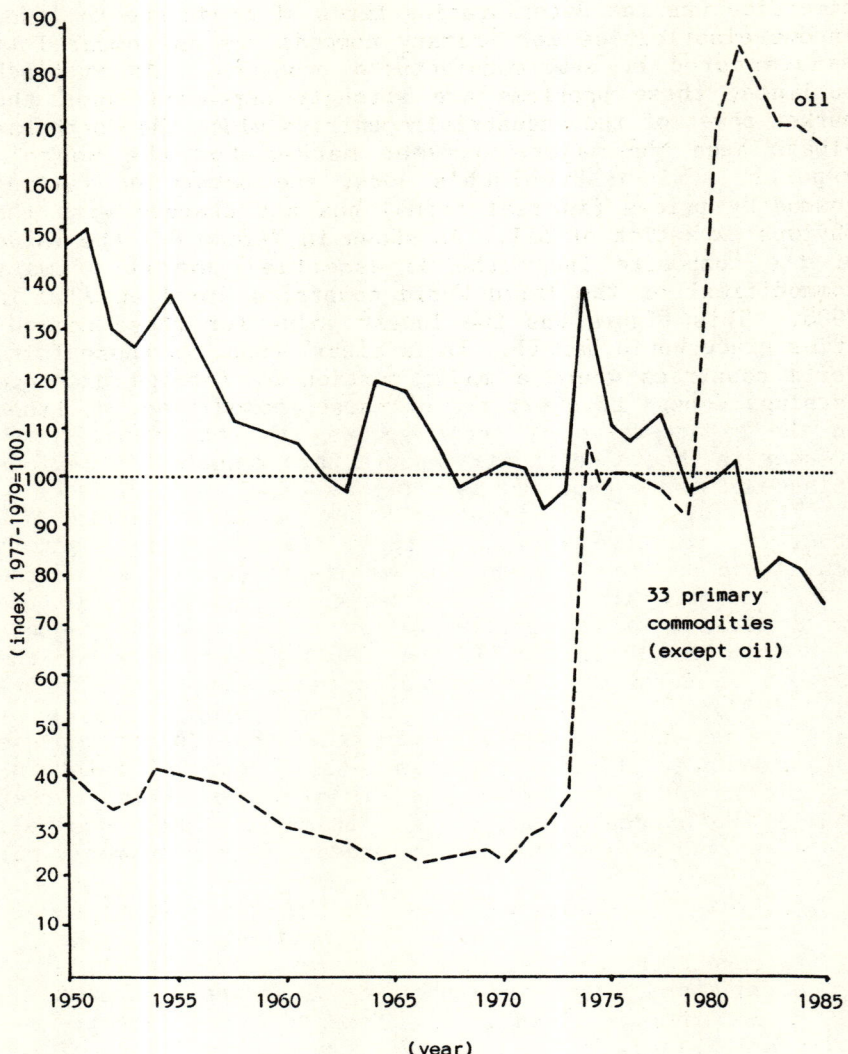

Figure 8.1 Composite index of commodity prices, 1950-1985

Source: World Bank. <u>Half-Yearly Revision of Commodity Price Forecasts</u>

In essence, the nonoil primary commodity exporting peripheral countries are faced with an acute dilemma. On the other hand, their foreign debt obligations require these countries to export more primary commodities (including foodstuff and cash crops) to meet foreign exchange earning targets. On the other hand, since these export efforts coincide with the period of declining per capita production in much of the Third World, the net result is further shortages in the domestic food supply.

In view of these problems it is debatable whether the international community could undertake measures to stabilize primary commodity prices. Since the GATT agreement in 1947, the North has maintained that market forces of supply and demand, free of state intervention, could determine the right value of commodities. This notion of free market determination of prices is largely a myth. Most often the role of governments include providing the confidence and rules to create markets as well as helping them function well in adjusting to supply and demand conditions (Lindlom, 1977). Whether we like it or not, the government intervention is a reality and control of prices are usually achieved through two mechanisms. The first mechanism is direct control through import tariffs of subsidies, setting of price ceilings, export control and taxes, and farm price supports. The second mechanism is indirect controls through exchange rate policies.

Over the years, the South has attempted to establish a united stand concerning their exports to the North. Faced with increasing dependency in North-South relations, the NIEO outlined strategies that would pave the way for stabilizing nonoil primary commodity prices. Encouraged by the success of OPEC in raising oil prices in 1973-1974, the other LDCs pushed for and acquired United Nations' approval to establish an Integrated Commodity Program (ICP) at the United Nations Conference on Trade and Development (UNCTAD IV) in Nairobi. This plan was perhaps the most ambitious design for controlling primary commodity prices in recent history. The integration in the ICP, for ten leading commodities[2], was to be provided by $6 billion Common Fund (CF) to finance the individual International Commodity Agreements (ICAs). In this plan the ICP emphasized the use of international buffer-stocks and controls on production or exports to avoid large scale fluctuations in commodity prices. In addition to these measures, the resolution called for improved information, improved market access, improved infrastructure, enlarged compensatory financing for export short falls, research and development (UNCTAD, 1976).

Despite the initial enthusiasm about the program, negotiations on the CF proceeded very slowly in the following years. Finally in June 1980 an agreement was reached on a much smaller CF, totaling $400 million, at voluntary basis. Furthermore, the ICP only resulted in ICAs for coffee, cocoa, and rubber. It is interesting to note that all of these commodities have pre-1976 ICAs that expired by the late 1970s. In other words, no ICA was reached on new agricultural commodities during the post-Nairobi years. Table 8.1 provides information concerning the ICAs on cocoa, coffee, sugar and rubber. In addition, fluctuations in the commodities' prices are shown in Figure 8.2. As shown in Table 8.1, control of two commodities's prices, cocoa and rubber, were to be based on the buffer stock arrangement, when the price of a commodity approaches the floor price, the buffer stock manager would enter the market and purchase sufficient amounts of the product to strengthen prices. Likewise, when the price approaches the ceiling-price, the manager would enter the market and sell enough quantities of the product to reduce prices. In comparison to buffer stocks, export quotas operated on quota limits of the main exporter. According to this procedure, if the world prices rise above a pre-determined limit, export quotas may be increased to lower the level of national stocks. This increase in export quotas would then stabilize prices rather than raise them.

Information in Table 8.1 and Figure 8.2 point out that among these ICAs only the coffee and rubber ICA is still effective. The sugar ICA's economics provision expired in December 1984 and no new agreement has been negotiated. The cocoa ICA expired 1986 and negotiations concerning its renewal were abandoned in the spring of the same year.

Fluctuations in the prices of these commodities also explain the problem encountered in these cases. Neither one of these commodities' prices were maintained within the agreed price ceilings. Simply, international market forces especially U.S. foreign agriculture policies, have contributed to this outcome. In both cases, the U.S. felt the target price range was too high and did not support the implementation of the ICA. Also, the sugar ICA had to compete with the European Community's shift from key importer to major exporter.

Table 8.1

Current International Commodity Agreements in Agriculture

Item	Cocoa	Coffee	Rubber	Sugar
Date of Current Agreement	1981[a]	1983	1980	1978[b]
Duration (years)	3	6	5	5
Extension (years)	2	---[c]	2	2
World Trade ($ billion/1984)	2.6	11.0	3.6	10.1
% from South	79	76	93	75
Dependency[d]	6	21	3	9
Principal instrument	buffer stock	export quota	buffer stock	export quota
Permitted price range (%)	±18	±15	±20	±43
Buffer stock as % of	16	---	15	---

[a] Expires September 1986; negotiations on renewal abandoned in Spring 1986.
[b] Economic provisions expired December 1984.
[c] Extended for an indefinite period.
[d] Number of countries, based on a sample of 83, in which the commodity accounted for more than 10% of exports in 1980.

Source: World Bank, World Development Report 1986, (New York: Oxford University Press), p. 135; reprinted with permission.

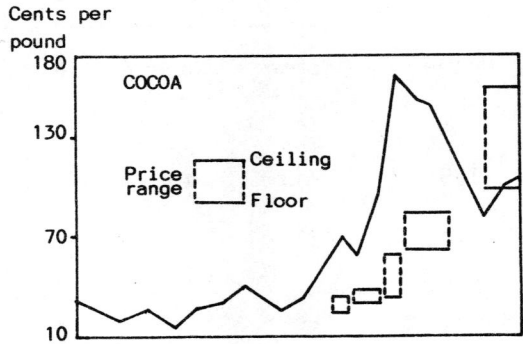

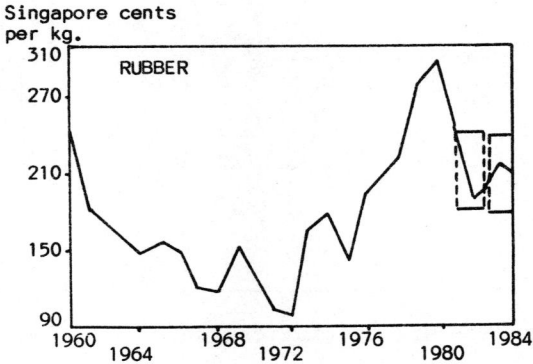

Figure 8.2 International Commodity Agreements, Prices, and Price Ranges

(Continued)

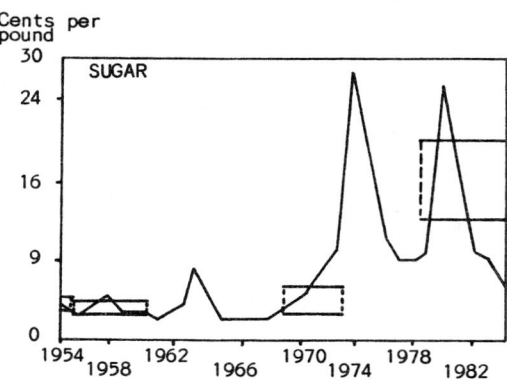

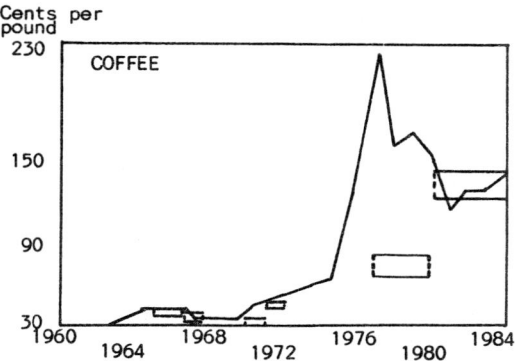

Figure 8.2

Source: World Bank, <u>World Development Report 1986</u> (New York: Oxford University Press), p. 136; reprinted with permission.

Among the two successful ICAs, that of rubber is more effective in maintaining a price within the agreed price range. The coffee ICA, on the other hand, has been less effective with prices remaining below the floor price. The major reasons for this trend in coffee prices have been the reluctance of the U.S. to support the agreement because it would have increased prices for the American consumers and the trade of large quantities of coffee outside the limits of the coffee ICA. The debt burden of the major coffee exporting Third World countries is one crucial factor behind these sales[3]. In their efforts to export more cash crops like coffee, the debtor countries actually contributed to the decline in commodity prices, undermining their status in North-South relations.

The above developments are indicative of the fact that the ICAs as a whole have not successfully achieved the goals set in the NIEO. The various commodity price agreements have proven difficult to manage and had to be advanced. Moreover, the once much championed ICP has been a total fiasco. As Taylor (1982) pointed out the major reason for the ICP's failure was the suspicion of the developed countries' concern over the Third World states' "hidden agenda" aimed at increasing the price of these commodities above their market equilibrium prices. At the present time, it is unrealistic to expect new agreements on an effective ICP in the near future. Even the UNCTAD in recent years seems to favor other commodity price controls, such as, processing primary commodities prior to exporting them and employing compensatory financing (UNCTAD, 1983). The latter option includes reliance on the IMF's Compensatory Financing Facility (CFF) and the European Community's STABEX[4]. Both options require large quantities of capital that would eventually come from the richer countries of the North. Since the North has been unwilling or unable to provide the necessary capital, the impact of compensatory financing on primary commodity exporters has fallen far short of expectations. Ironically, this resembles the very problem the Common Fund faced in the early 1980s.

PROBLEMS IN AGRICULTURAL TRADE

In order to import the necessary inputs for agricultural and non-agricultural development, Third World countries are in constant need of essential foreign hard currencies. Generally these foreign currencies come from the following available sources: (i) export earnings, (ii) foreign aid,

(iii) direct foreign investment, (iv) external loans, and (v) selling domestic assets to foreigners. Among these options only exports provide a safe route for obtaining much needed foreign currency. As discussed in the next section, the flow of official assistance and direct foreign investment to peripheral countries has declined since the early 1970s. Furthermore, external loans have replaced these as primary sources of foreign currency, thus causing devastating problems for the Third World. The sales of domestic assets to foreigners is not an attractive option to developing countries' governments, though some have considered this idea over the years. Therefore, it is essential for a developing country to acquire foreign hard currency through exports.

Since agriculture is the largest sector for most Third World states' economies, exports will have to come from this sector---at least initially. This is a very important responsibility for the agricultural sector, for if it fails to provide a trade surplus, the result may be delayed investments on domestic development projects. Despite this crucial rule, agriculture in Third World countries has failed to generate the much needed trade surplus since the mid-1970s as shown in Figure 8.3.

Region specific figures provide some significant observations. With the exception of the Asian market-oriented LDCs, the other regions have experienced deep trade deficits in agriculture. The most alarming case is that of Subsaharan states where there was a net deficit of 160 index points in 1983. The second largest deficit is observed in the Middle East and the third is in Latin America. It is interesting to note that during the first half of the 1970s, all of these regions were experiencing agricultural trade surpluses or deficits that were minimal. Their problem seems to have started in 1977-1979 and later exacerbated to alarming rates. However, these states' figures do not show significant trade surplus either. In comparison to the LDCs' recent agricultural trade deficits, the developed Western countries show increased surpluses in this area. Several factors serve as causes. First is the continuing decline in per capita agricultural production in most Third World countries. At times when the population growth rate surpasses per capita food and agriculture production less goods become available for exports. The second problem is that of falling primary commodity prices discussed in the previous section. And third is the protectionist trade policy of the industrial West. In addition to these factors there are country specific domestic policies in LDCs that inhibit

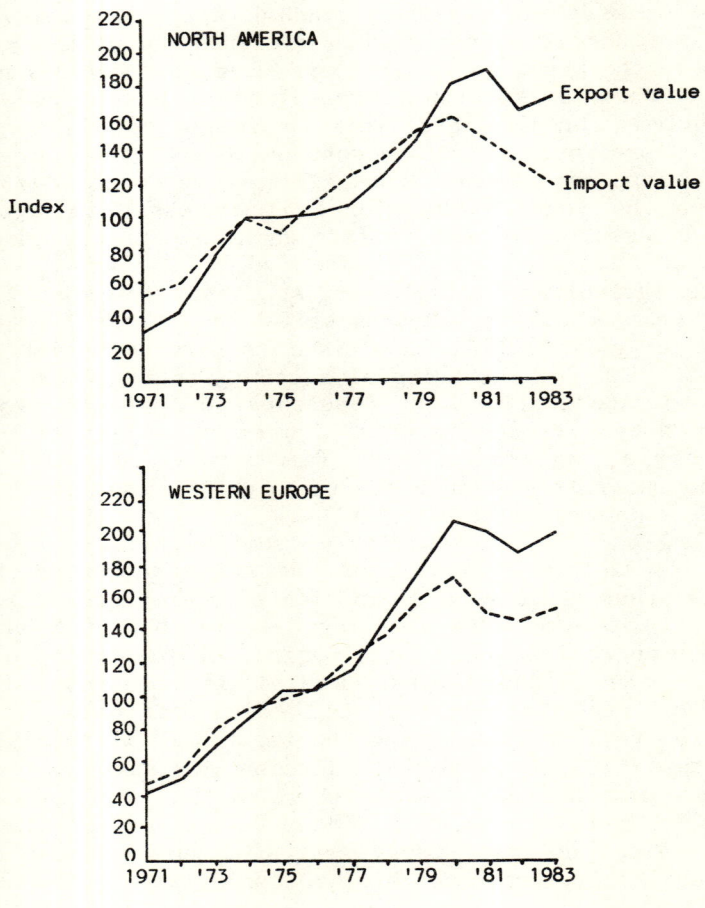

Figure 8.3 Agricultural Trade Index by Region (1974-1976 = 100)

(Continued)

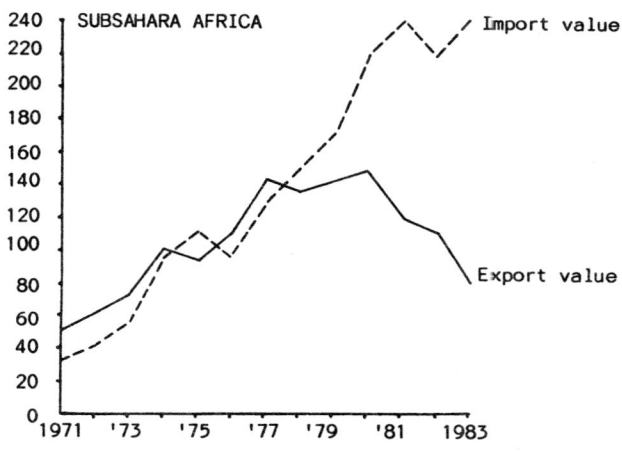

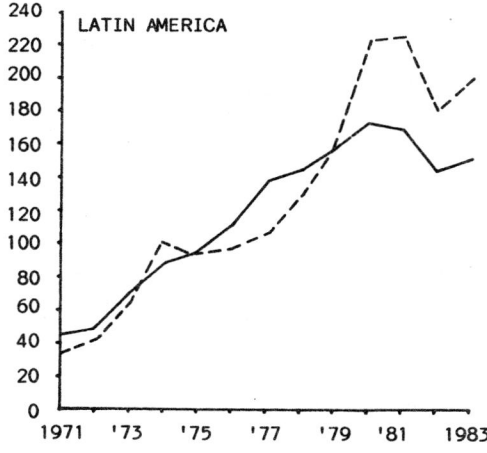

(Continued)

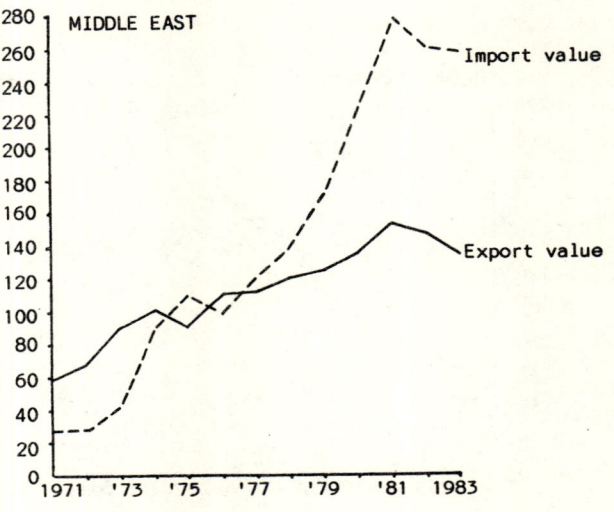

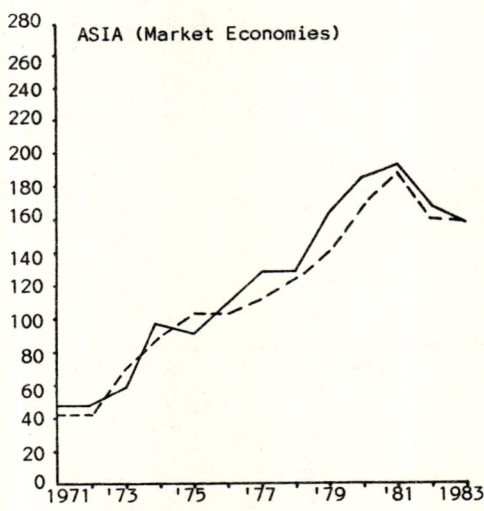

Source: Weighted averages estimated from country figures in the FAO Trade Yearbooks (selected years).

agricultural exports. One such practice is the taxation policy on agricultural exports. On such practice is the taxation policy on agricultural exports that is aimed at generating revenues for the poorest states. One major consequence of these export taxes is a significant squeeze on export activities in agriculture, resulting in reduced investment in these activities. Likewise, artificial exchange rate policies of the government play a role in inhibiting agricultural exports. For the purposes of this chapter, I will focus on trade relations between the North and South to show some of the unfair trade policies that limit peripheral countries' export of agricultural commodities.

While most of the industrializing states of the Third World practice import-substitution policies against imported manufactured goods, the LDCs face similar protectionist measure for their agricultural exports to the industrial countries. Given the internationalization of the world economy and the crucial role the industrial states' trade policies play in the world trade, the future of most Third World countries' agriculture depends increasingly on these international markets, over which they have very little control. In general, the protectionist agricultural trade policies of the Northern states cause world prices to be lower than they would be with less protectionism. Valdes and Zeitz (1980) have shown that a 50 percent reduction in OECD trade barriers would have increased agricultural exports of 56 LDCs by 11 percent or $3 billion (in 1977 dollars). Several crucial reasons can be found for the protectionist measures of the North: (i) many countries consider the protection of domestic agriculture as vital to their economy; (ii) lobby efforts by domestic farmer groups on the government---in the U.S. where Congressmen are elected every two years, this is a major concern for those who run from farm states; and (iii) protectionist policies prevent disruption of the domestic price market. Generally, the programs adopted operate through pricing objectives which tend to be higher than the cost of imports.

Among protectionist trade policies fixed tariffs are rarely practiced because such measures cannot stabilize domestic prices since they tend to fluctuate along the world prices. Import quotas, on the other hand, are extensively employed by the U.S, Canada, the European Community, Japan, and Switzerland. Their purpose is to isolate the domestic market from variations in the world market and often result in raising domestic prices. Sometimes, such import quotas are found as "voluntary-export-restraint" agreements between importers and exporters.

Another protectionist agriculture trade policy is the "import levies". This is the major aspect of the European Community's Common Agriculture Policy (CAP), though import levies are also practiced by Austria, Sweden, Switzerland, and the U.S. (for sugar trade). Import levies make up the difference between the price of imports and their officially determined fixed entry prices. This raises the duty-paid price of imported agricultural commodities slightly above domestic prices and the Western consumers buy goods whose prices have artificially risen. As a result imports are limited and import competition is minimized. On the other side of the production-consumption relationship for the producers in the South this means continuation of export earning below their real market values. The World Bank (1986: 117) statistics on protectionist trade measures in 1984 show that application of various nontariff barriers in industrial countries varied from 6.6 percent for tea, coffee, and cocoa to 70 percent for sugar and confectionery. The number of import items subject to the nontariff barriers (quotas, quantitive restrictions, and variable levies) as a percentage of the total number of import items are presented in Table 8.2.

These figures show that all of the three trade barriers have been extensively employed by the industrial states. It is important to note that while the frequency of applied trade barriers for all agricultural products was 29.7 percent in 1984, this figure was a mere 9.4 percent for manufactured goods. Furthermore, among the agricultural goods the main cash crops (tea, coffee, and cocoa) experienced the lowest trade barriers. After all, these commodities which are in constant demand in the Western consumer markets have to be imported and do not threaten any domestic Western producers. These figures highlight the irony in the industrial states' policy suggestions at promoting free trade principles of GATT in agricultural trade. It seems that the North's fervent advocacy of laissez-faire policies in international agricultural markets (World Bank, 1986: Ch.8; Aho and Bayond, 1985; OECD, 1982; McDonald, 1981: 362-366) is nothing more than a mere lip service to GATT principles. As indicated in this section, agricultural policies of industrial states are plagued with departures for laissez-faire economics. The overall impact of these protectionists measures in the North towards the agricultural exports of the South is continuous trade earnings by peripheral countries at a level below their market potential. This in turn is a major complicating factor in these states' attempts to meet their foreign obligations and finding additional sources of income to promote domestic economic growth.

Table 8.2

The Frequency of Application of Various Nontariff Barriers in Industrial Countries, 1984 (percent).[a]

Commodity	Tariff Quotas	Quantitative Restrictions	Variable Levies	Total[b]
Meat & live animals	12.3	41.0	23.8	52.2
Dairy products	6.9	29.6	25.6	54.6
Fruits & Vegetables	15.7	18.8	0.8	33.1
Sugar & confectionery	0.0	21.7	58.0	70.0
Cereals	1.7	10.9	21.7	29.0
Other food	0.8	16.3	13.2	27.0
Tea, coffee, cocoa	0.4	4.0	2.5	6.6
Other beverages	18.5	22.9	0.3	7.8
All agriculture	8.2	17.2	8.2	29.7
All manufactures	2.2	6.7	0.0	9.4

[a] The industrial countries are Australia, the European Community, Finland, Japan, Norway, Switzerland, and the U.S.
[b] This number can be less than the sum of other columns if some imports are subject to more than one barrier.

Source: World Bank, World Development Report 1986, (New York: Oxford University Press), p. 117; reprinted with permission.

IMPACT OF THE INTERNATIONAL DEBT CRISIS ON THIRD WORLD ECONOMIES

The current international debt crisis facing many peripheral countries relates directly to the balance-of-payment and investment financing policies of these states in conjunction with the industrial West's fiscal and monetary policies aimed at combatting recession. Ordinarily, domestic investments can be financed through domestic savings and export earnings, provided that there is ample trade surplus. When such resources are insufficient to meet investment targets, governments rely on external sources such as official aid, direct foreign investment, and loans to meet their investment needs.

The impact of the 1974-1975 recession on export earnings of peripheral states proved to be disastrous. Slower economic growth in the core with heightened protectionism against imports reduced the volume of growth of world trade from about 9 percent per year between 1965 and 1973 to about 3.9 percent per year between 1973 and 1978 (FAO, 1973-1984a). During this period, the decline in the growth of periphery country exports was from 6.4 percent to 3.6 percent per year. In fact, when one examines these countries' share in merchandise trade, the decline becomes more vivid. The peripheral states' exports accounted for 24 percent of total merchandise trade in 1960. In retrospect, the World Bank (1979; 5) reports this figure as 21 percent in 1976. In the area of nonoil primary goods, the decline was from 37 percent to 34 percent for the same period. The only sector where the developing countries' share of world exports increased was manufactured products, form 6 percent in 1960 to 10 percent in 1976. However, despite this increase, the growth rate of manufactured exports slowed down from an average of 15 percent per year in the period 1964-1973 to about 10 percent per year during 1974-1977. This net decline in exports for the peripheral countries provoked increased balance-of-payment problems shown in Table 8.3.

It is apparent form these figures that the peripheral countries, with the exception of high income oil exporters, faced serious deficits during 1973-1976. Low income African states and the middle income oil importers suffered the most, mainly due to their reduced exports and the increased price of imported oil. The crisis briefly subsided during 1976-1979 when the market entered its recovery phase. However, with the advent of the second oil shock in 1979 and the latest market recession of 1980-1983, the developing

Table 8.3

Current account balance as percentage of GNP in peripheral countries
(billions of U.S. dollars)

Country Group	1970	1973	1974	1975	1976	1977	1978	1979	1980	1981	1982	1983	1984
All Periphery	-2.3	-0.9	-1.9	-3.9	-2.2	-2.2	-2.6	-2.0	-2.3	-3.9	-3.7	-2.8	-1.8
Low Income													
Asia	-0.9	-0.6	-7.8	-1.2	-0.1	0.4	-0.1	-0.6	-1.4	-0.5	0.2	-0.2	-0.6
Africa	-3.9	-4.4	-7.8	-10.2	-7.3	-7.6	-8.3	-7.7	-9.8	-10.5	-12.0	-10.0	-9.4
Middle Income Oil													
Importers	-3.2	-0.8	-4.4	-5.3	-2.9	-2.3	-2.2	-3.2	-4.1	-5.2	-4.7	-4.4	-2.7
Exporters[a]	-3.2	-0.9	-5.7	-5.5	-2.8	-1.6	-1.7	-3.1	-3.6	-4.2	-4.0	-3.1	-1.3
Middle Income Oil													
Exporters	-3.0	-1.1	3.3	-3.4	-2.4	-3.6	-5.1	-0.2	0.8	-3.8	-4.4	-2.1	-0.7
High Income Oil													
Exporters	15.7	21.2	51.5	40.2	35.0	26.3	15.5	21.2	31.4	32.2	20.1	-4.7	--

[a] oil importing middle income countries that are major exporters of manufactured goods.

Sources: Estimated from country account figures in World Bank, World Tables 1980 and 1983 and World Development Report 1985 (New York: Oxford University Press, 1985), Table 14.

countries again experienced serious balance-of-payment deficits. The current account balances of the peripheral states are also closely related to their domestic savings difficulties. As mentioned above, domestic savings provide one crucial source of capital for investments. Table 8.4 presents estimates of the share of domestic savings and gross domestic investments in GDP for these countries.

These figures show that while gross domestic investments in all country groups increased since 1971, domestic savings have not followed suit with the exception of the low-income Asian countries, largely China and India. In these two countries the ratio of domestic savings to domestic investments has been close to 100 percent in recent years. However, in all other regions, the impact of recession has reduced this ratio with low-income African states showing the worst performance.

Thus, the peripheral states have increasingly relied on external sources to finance their development needs. As late as 1972, external financing to the developing countries came from non-debt-crediting flows such as official assistance and long-term direct foreign investments. However, with the impact of the first oil price shock on the industrial countries' economies, private financial markets significantly expanded. As the oil-rich OPEC states began to invest their petrodollars in the financial markets of the core, large private banks became the single most important channel for transfer of capital to the peripheral states. Two factors contributed to this development. First, with the quadrupling of oil prices in 1973, official assistance to the developing countries declined in real terms. Instead of working toward reaching the 0.7 percent of GNP for official assistance, as the United Nations recommended, the contribution of the major industrial states declined during 1970-1980 (OECD, 1983; 10-15). The second factor was the reduction in the flow of direct foreign investment into the peripheral states by 1981. Estimation of the share of private direct investments in the total net source receipts of the Third World from all sources in real terms shows the following trend: 21.2 percent-1975, 14.9 percent-1977, 14.6 percent-1979, and 14.8 percent-1982 (OECD, 1983; 52).

Faced with declines in these capital flows, the peripheral states turned toward private banks for credit in order to maintain high investment levels in their economies. This increase in demand for credit generated by ambitious development plans and the sharp rise in oil prices was more than steadily matched by the inflow of petrodollars into the market. Credits were readily available, spreads were

Table 8.4
Gross domestic savings and investment in peripheral countries (weighted averages)

Country Group	Gross Domestic Savings (A) as percent GDP				Gross Domestic Investment (B) as percent GDP				A/B (percent)			
	1971-73	1974-75	1976-79	1980-82	1971-73	1974-75	1976-79	1980-82	1971-73	1974-75	1976-79	1980-82
All Periphery	14	15	19	18	20	22	24	25	70	68	79	72
Low Income Asia	13	16	20	19	16	19	23	22	81	84	87	86
Low Income Africa	9	10	11	6	14	16	17	13	64	62	65	46
Middle Income Oil Importers	14	15	19	19	20	23	26	27	70	65	73	70
Middle Income Exporters of Manufactures	17	19	23	22	22	25	27	28	77	76	85	79

Sources: Estimated from country accounts in World Bank, World Tables 1980 and 1983, (Baltimore: The Johns Hopkins University Press, 1980 and 1983).

generally falling until mid-1979, and interest rates were barely positive which created conditions that favored borrowers. With such conditions, the peripheral countries accumulated a foreign debt of $547 billion, disbursed and undisbursed, by the end of 1982. Table 8.5 presents the distribution of external debt by country groups. The middle income exporters of manufactured products followed by the middle income oil exporters accumulated the largest debt. Furthermore, for these country groups, some 78.6 percent and 71.8 percent of their total debts respectively were owed to private financial institutions. In comparison, the private source total debt ratio was 13.6 percent in low-income Asia and 26.9 percent in low-income Africa. Among the middle income states, rank of liabilities were: Mexico, $84.1 billion; Brazil, $62.7 billion; South Korea, $37.3 billion; Argentina, $28.2 billion; Turkey, $22.6 billion; and Egypt, $21.8 billion. Since 1982 these figures have increased significantly as the total debt liabilities of peripheral states reached $687 billion by the end of 1984.

The favorable market conditions that led to the accumulation of huge foreign debts were not, however, without their danger. First, commercial bank loans are characterized by higher interest rates than official source loans. Furthermore, those commercial loans have shorter maturity. In the late 1970s the large accumulation of loans from private sources resulted in a disproportionate increase in the debt-service ratio of the borrowers as shown in Table 8.5. By 1982 the debt-service ratio of peripheral states had reached the 20.5 percent mark, seriously hurting the export earnings of these states. The heaviest burden was among the middle income countries. World Bank (1985: 181-183) data show that the debt-service ratio for key individual states in 1982 was: Mexico, 33 percent; Brazil, 43 percent; Argentina, 25 percent; South Korea, 14 percent; Turkey, 19 percent; and Egypt, 21 percent.

The second danger was more closely related to the fiscal and monetary policies of the industrial market states. When the second oil price shock hit and the industrial states adopted tight monetary policies, the cost of borrowing skyrocketed. Prior to 1979 the cost of borrowing fluctuated between 6.5-7.75 percent as estimated by LIBOR (London Interbank Offer Rate) plus average spread. After allowing for inflation, this meant that real interest rates increased from less than 3 percent to more that 8 percent. The final outcome was the debt crisis that emerged in August 1982 when Mexico announced that it could not service her debt. Soon after Mexico, Argentina and Brazil declared their inabili-

Table 8.5
Debt indicators for peripheral countries (ratios in percent and amounts in billions of U.S. dollars)

Country Group and Item		1970	1974	1976	1978	1980	1981	1982	1983	1984
All Periphery Countries:	Total Foreign Debt	69	141	204	314	430	488	547	619	687
	Debt Service Ratio[a]	14.7	11.8	13.6	18.4	16.0	17.6	20.5	19.0	20.0
Low Income Asia:	Total Foreign Debt	12	18	22	29	38	40	43	46	53
	Debt Service Ratio	12.4	7.8	7.7	7.2	8.0	9.3	10.9	8.3	8.4
Low Income Africa:	Total Foreign Debt	3	7	10	15	21	23	25	25	27
	Debt Service Ratio	6.1	8.6	8.6	9.6	12.5	13.8	15.7	16.5	19.9
Middle Income Oil Importers:	Total Foreign Debt	12	21	27	43	68	79	89	98	108
	Debt Service Ratio	13.6	11.4	14.8	20.9	17.2	20.8	22.7	23.1	24.9
Middle Income Exporters of Manufactures:	Total Foreign Debt	24	57	82	124	167	191	216	242	267
	Debt Service Ratio	15.1	13.7	14.2	17.7	16.1	17.1	19.3	16.2	16.0
Middle Income Oil Exporters	Total Foreign Debt	18	38	63	103	136	155	174	208	232
	Debt Service Ratio	18.1	11.0	14.5	22.9	17.8	19.8	25.0	26.1	28.1

[a] debt service ratio is the ratio of the sum of interest payments and repayments of principal on external public and publicly guaranteed debt to total export earnings.

Source: World Bank, World Development Report 1985 (New York: Oxford University Press, 1985), p.24; reprinted with permission.

ties to finance their debt service payments. Perhaps the most crucial outcome emerging from this crisis was the increasing power of the IMF to dictate the future economic policies of the debtor countries.

GROWING INVOLVEMENT OF THE IMF IN DOMESTIC POLICIES OF THIRD WORLD STATES

At the macroeconomic policy level, between 1973 and 1984, peripheral states gradually increased their adoption of the IMF's monetarist stabilization programs in order to receive the Fund's approval for debt rescheduling and additional financial assistance from private sources. In the 1980s it had generally become a rule that without the IMF's certificate of approval, funds from other sources would dry up. The general framework of various IMF policies extended to debtor states do not vary significantly and emphasize the following policy suggestions:

1. Elimination of the disequilibria in the monetary and fiscal spheres as well as in prices and the foreign sector,
2. Reduction of the influence upon the new economic and political system of the supposedly antagonistic workers groups by modifying or abolishing labor legislation and by holding down real wages,
3. Implementation of a new long-term development model, characterized by a more export-oriented growth, by the development of private financial markets and by a reduced government role,
4. Opening of economy for world and capital flows,
5. Redistribution of resources toward the private capitalist sector with the aim that this sector should dynamize growth, based on market principles. (Kirkpatrick and Onis, 1985: 347-350).

The increasing adoption of IMF austerity measures by the peripheral states is directly related to the magnitude of the debt crisis facing these countries. While during the 1960s and 1970s no more than five countries required debt rescheduling by the Fund in a single year, 13 states asked for assistance in 1981, 10 in 1982, 21 in 1983, and 34 in 1984. In dollar terms these rescheduling translate as $373 million in 1975, $6.2 billion in 1979, $3.7 billion in 1980, $5.8 billion in 1981, $2.4 billion in 1982, $51.0 billion in 1983, and $116.2 billion in 1984 (World Bank, 1985: 28).

The geographical distribution of the rescheduling by number of countries according to their geographical distribution are presented in Table 8.6.

These figures show that while Latin America debt rescheduling accounted for the largest amount in dollar terms ($152,680) during 1975-1984, Africa surpassed all other regions in the number of countries seeking debt reschedulings. The breakdown of the number of countries by region is: Africa, 17 countries for 46 cases; Latin America, 15 countries for 30 cases; and Asia, 3 countries for 5 cases. Since 1975, more African states were faced with debt serving problems than in any other region. Often world attention is focused on Latin America where the largest value of debt is found. However, if one examines debt figures of countries as percentage of their GNPs, it becomes apparent that the African states match with those in Latin America as being the worst hit by the debt problem. World Bank (1985: 204-205) statistics show that in 1984 the weighted average for low-income Subsaharan countries for debt as percentage of GNP was 52.3 percent and for the middle-income Subsaharan countries the figure was 29.0 percent. In comparison Brazil had 29.3 percent, Mexico-49.1 percent, Argentina-32.1 percent, Venezuela-19.8 percent, and Chile-39.3 percent.

A close examination of the IMF austerity measures adopted by Third World countries reveals important designs, all aimed at the further integration of peripheral countries into the world market. In this process the one sector that occupies the driver's seat is the domestic private sector in close cooperation with foreign capitalists. In contrast, the role of the state in economic development is deemphasized.

Among these measures, tight monetary policies and restraint of real wages, through maintenance of current wage hikes below the level of inflation, aim at reducing demand in the domestic market. This, in turn, is expected to help reduce inflation, stabilize growth, and make more goods available for export. Often this emphasis on export promotion translates into increases in agricultural exports in the form of cash crops which require more domestic investment emphasis on this sector rather than in the production of agricultural products geared to feeding the growing local population.

The reduction of real wages is further aggravated when governments decide to eliminate state subsidies on essential food items in order to allow market forces to operate. In theoretical terms this is a rational decision because lift-

Table 8.6
Distribution of debt reschedulings by region, 1975-1984
(number of countries/ billion $U.S.)

Region	1975	1976	1977	1978	1979	1980	1981	1982	1983	1984
Latin America	1 0.2	1 0.9	0 0.0	1 0.5	3 0.9	1 0.6	3 0.7	2 0.1	8 42.1	10 104.1
Africa	0 0.0	1 0.2	2 0.2	1 0.1	3 1.7	4 0.5	8 1.7	6 0.4	11 5.7	10 2.3
Asia	1 0.2	1 0.2	1 0.1	0 0.0	0 0.0	0 0.0	1 0.3	0 0.0	0 0.0	1 5.6
Other	0 0.0	0 0.0	0 0.0	1 1.2	2 3.6	1 2.6	1 3.1	2 1.8	2 3.3	13 4.2
Total	2 0.4	3 1.3	3 0.3	3 1.8	8 6.2	6 3.7	13 5.8	10 2.3	21 51.1	34 116.2

Source: Calculated from country figures in World Bank, World Development Report 1985 (New York: Oxford University Press, 1985), p. 28.

ing of government subsidies would help reduce the domestic deficit. However, in countries where a substantial portion of the population depends on these subsidies for survival, such decisions are bound to create domestic unrest. In fact, such decisions resulted in immediate "bread riots" in Egypt, Sudan, and Tunisia and forced the governments and the IMF to reexamine their policies and reintroduce subsidies.

The growing emphasis on stabilization programs have also resulted in reduced government expenditure in social programs on health care and education. An estimation of weighted averages for government expenditures for countries affected by IMF packages during 1973-1982, from the data available in the World Tables, provide the following results. In Latin America (n=15) educational expenditures declined by 17.3 percent and health care spending fell by 18.7 percent. In East Asia (n=8) the declines were 9.1 percent and 11.2 percent respectively. The figures for Middle East (n=8) also show reductions by 20.3 and 18.4 percent for these sectors. The worst declines were in Subsaharan Africa (n=25) where educational expenditures fell by 32.5 percent and health care spending fell by 40.3 percent. These reductions in government spending on social programs coupled with reduced real wages present the social costs of the austerity measures that are designed to stabilize troubled economies.

Another aspect of the IMF packages is the call for export-led growth policies that have changed the nature of the dependency in the world market. What export-led growth has meant for peripheral states is the adoption of export promotion policies for domestic industries, promotion of cash crop production, and the establishment of Free Export Processing Zones (FEPZs) for investments by multinational corporations (MNCs). Earlier examples of FEPZs were observed in Mexico, South Korea, Taiwan, Hong Kong, and Singapore during the late 1960s. During the 1970s, FEPZs spread to all regions of the world. The OECD reports that in 1970 eleven such zones existed around the world and that this number increased to 93 by 1981 (Germidis, 1984: 22).

In essence, the countries that have adopted FEPZs have become production centers for manufactured goods and processed agricultural products exported to the consumer markets of the industrial economies of the West. The expansion of this form of production can be identified with the new international division of labor by world capitalism where production of commodities is localized in specific regions of the labor-surplus periphery. This form of export promotion is highly dependent of the inflow of external capital and presents a more sensitive form of dependency between the periphery and core states. Export expansion on these sites depends on the constant growth of consumer markets in the industrial core which, in turn, is dependent on the fiscal and monetary policies of the industrial Western states.

Hence, when the export-led policies of the Fund are evaluated in their overall impact on Third World development, it becomes apparent that these policies have been helpful for the exporter of manufactured goods. As explained in the previous section, the export of manufactures from the Third World countries was also aided by the lower frequency of nontariff barriers applied to these goods in the industrial states. However, the impact of these policies on the export of primary commodities have been negative during the post-1983 period. The post-recession years of the world market have been characterized by falling primary commodity prices. Furthermore, at time of low short-run price elasticity of demand for agricultural products in the industrial states, such declines in agricultural commodity prices have also aggravated the depressed export revenues of the debtor peripheral countries.

CONCLUSIONS

This study has demonstrated that agricultural problems facing peripheral countries are heavily dependent upon international factors over which they have almost no control. It is clear that various protectionist trade measures in the industrial states, artificially determined commodity prices, the international debt crisis, and the increasing intervention of the IMF in policy decisions of peripheral countries have aggravated the existing problem in the South. In this regard, the position of the peripheral countries' agricultural exports has not improved in the North-South relations since the first UNCTAD.

Today most Third World states face heavy debt servicing requirements. Their ability to finance this debt, while still investing for the future development of their economies, depends on their expanding export earnings. Yet, the export earnings for much of the Third World seem to have declined in the last decade. In this regard, these states that largely rely on agricultural exports have been among the hardest hit. With ever present trade barriers in the industrial states, these peripheral countries need to find other export commodities to generate external earnings. As shown in the case studies in Part I, several states have increased the production of cash crops, like coffee, tea cocoa, ground nuts, etc., often at the expense of food production for their domestic markets, in order to earn more foreign exchange.

This concern for the acquisition of foreign exchange at all costs presents serious problems for agrarian reform in Third World states. These countries are essentially faced with a catch-22 that results in "Agrarian Reform in Reverse". On the one hand, they need to expand their exports of agricultural goods to acquire hard currency to pay back their debts and to invest in domestic development project. On the other hand, the industrial market countries' trade policies and their push for the increased production of cash crops threatens the very existence of the Third World states' traditional food base.

The examination of nontariff trade barriers in the industrial West for agricultural imports indicated that cash crops, as a category, are subject to the lowest levels of trade restrictions. The frequency at which cash crops are subjected to nontariff trade barriers was 6.6 percent in 1984 as compared to 52.2 percent for meat products, 54.6 percent for dairy products and 29.0 percent for cereals. Interestingly enough, the exporters of manufactures where the frequency of applied nontariff barriers remain very low, do not confront these problems. This is a natural outcome of the expanding production base of the industrial West's multinational corporations. With so many MNCs producing goods in the peripheral states, destined for the markets of the industrial West, it is no accident that nontariff and tariff barriers for manufactures in the developed countries remain much lower than those barriers targeted for agricultural goods.

The bottom line of the present international atmosphere for the production and trade of agricultural commodities is that laissez-faire policies in this sector are not present and that market determination of primary commodity prices is still a myth. Agricultural development of the peripheral countries face many obstacles for growth that the numerous UNCTADs have been unable to solve. As long as such obstacles remain with us, we should expect future problems in the peripheral economies' agricultural sectors.

NOTES

1. The primary commodities of major importance to Third World states are: coffee, tea, maize, rice, wheat, sorghum, soybeans, ground nuts, palm oil, copra, ground nut oil, soybean meal, sugar, beef, bananas, oranges, cotton, jute, rubber, tobacco, logs, copper, tin, nickle, bauxite, aluminum, iron ore, manganese ore, lead, zinc, and phosphate rock.

2. These ten leading nonoil primary commodities are: coffee, cocoa, copper, cotton and cotton yarns, hard fibers and products, jute and jute products, rubber, sugar, tea, and tin.

3. The debt-service ratio of the key coffee exporting developing countries at the end of 1984 was: Brazil/26.0 percent, Columbia/24.4 percent, Costa Rica/25.3 percent, El Salvador/17.2 percent, Guatemala/16.5 percent, Honduras/15.2 percent, Mexico/34.3 percent, Cameroon/10.9 percent, Ethiopia/16.8percent, Ivory Coast/21.3 percent, Kenya/21.5 percent, and Zaire/9.0 percent. (World Bank 1986: Table 18).

4. The purpose of the IMF's CFF is to provide financial assistance to the IMF members experiencing balance of payment problems resulting from export shortfalls that are temporary. Its compensation rate is 62 percent and has been employed 112 times between 1977-1982, totalling transactions worth $19.2 billion. The interest rate on such drawings is IMF standard on a 3-5 year repayment schedule. The grant element in these drawings is about 20 percent. The STABEX, on the other hand, is restricted to the EC's African, Carribean, and Pacific (ACP) states. Exports of 48 agricultural products are covered in this arrangement. STABEX has no balance of payment difficulty requirements for drawings by individual countries. Between 1977-1982 there were 171 drawings by individual countries.

REFERENCES

Aho, C. Michael and Thomas O. Bayard (1985). "The 1980's: Twilight of the Open Trading System," In John Adams. ed. The Contemporary International Economy. New York: St. Martin's Press.

Avromovic, Dragoslav (1986). "Depression of Export Commodity Prices of Developing Countries: What Can be Done?" Third World Quarterly 8, 3 (July): 953-977.

Bagu, Sergio (1949). Economia de la Sociedad Colonial: Ensayo De Historia Comparado de America Latina. Buenos Aires: Liberia "El Ateneo".

_____ (1960). Argentina en el Mundo. Buenos Aires: Fundo De Cultura Economica.

Caporaso, J.A. (1978). "Dependence and Dependency in the Global System." International Organization (special issue) 32 (Winter).

Cardoso, Fernando H. (1962). Capitalismo e Escarvidao no Brasil Mardinal: O Negro no Sociedade do Rio Grandedo Sul. San Paulo: Difusao Europeia do Liyro.

Dos Santos, Theotonio (1967). El Nuevo Caracter de la Dependencia. No. 6, Santiago, Chile: Cuadermos de Estudios Socioeconomicos.

Food and Agriculture Organization (1973-1984a). Trade Yearbook. Rome: FAO.

_____ (1973-1984b). Production Yearbook. Rome: FAO.

Frank, Andre Gunder (1966). "The Development of Underdevelopment." Monthly Review 18 (September): 17-31.

Furtado, Celso (1964). Development and Underdevelopment. Translated by Ricardo W. De Aquiar and Eric C. Drysdale. Berkeley: University of California Press.

Germidis, Antoine Basile ef Dimitri (1984). Investing in Free Export Processing Zones. Paris: OECD Publications.

Kirkpatrick, Colin and Ziya Onis (1985). "Industrialization as a Structural Determinant of Inflation Performance in IMF Stabilization Programs in Less Developed Countries Journal of Development Studies 21, 3 (April): 347-361.

Lindblom. Charles (1977). Politics and Markets. New York: Basic Books.

McDonald, Alonzo L (1981). "Implications of the Multilateral Trade Negotiations for Agricultural Trade," In Richard G. Woods. ed. Future Dimensions of World Food and Population. Boulder: Westview Press.

OECD (1982). Problems of Agricultural Trade. Paris: OECD Publications.

Prebisch, Raul (1959). "Commercial Policy in the Underdeveloped Countries." American Economic Review 49: 251-273.

Taylor, Lance (1982). "Back to Basics: Theory for the Rhetoric in the North-South Round." World Development 10, 4 (April): 327-336.

UNCTAD (1964). Towards a New Trade Policy for Development. E/Conf. 46/3. New York: United Nations.

_____ (1974). Resolution of the United Nations Conference on Trade and Development: Integrated Programme for Commodities. TD/RES/93 (IV. Nairobi).

_____ (1983). Commodity Issues: A Review and Proposal for Further Action. TD/273. Geneva.

World Bank (1979). World Development Report. New York: Oxford University Press.

_____ (1980). World Tables. Baltimore: John Hopkins University Press.

_____ (1983). World Tables. Baltimore: John Hopkins University Press.

_____ (1985). World Development Report. New York: Oxford University Press.

_____ (1986). <u>World Development Report</u>. New York: Oxford University Press.

_____ (selected years) <u>Half-Yearly Revision of Commodity Price Forecasts</u>. Washington: World Bank Publications.

Yeshilada, Birol (1987). "World Market Recession and Economic Development: A Look at Alternative Rescue Policies for the Third World." <u>International Journal of Contemporary Sociology</u> Vol. 24, no. 1&2 (January-April): Article #6, pp. 1-20.

9

Political Implications of International Monetary Fund Conditionality for Latin America

Adalberto J. Pinelo

INTRODUCTION

During the past three decades, the International Monetary Fund (IMF) has become so powerful an agency in the Third World, and in Latin America in particular, that we need to reappraise its role and importance. It is the thesis of this paper that the IMF has drastically altered the realities of internal Latin American politics and that it has also redefined U.S.-Latin America relations. More precisely, the IMF has significantly eroded the capacity of the region's governments to make independent policy decisions and has narrowed the parameters within which Hemispheric nations must operate.

The IMF represents not only a traditional actor with an expanding role. It also represents a new vehicle for the exercise of U.S. influence in Latin America, one that performs far more effectively than other, more traditional, bilateral government-to-government dealings.

Writing in the late 1960s, Robert Gilpin characterized the relations of the Soviet Union with the members of its bloc as a system "held together largely through the exercise of Soviet military power..." By the way of contrast he said that "economic relations have been an important cement building the American bloc together" (Gilpin in Modelski, 1979: 75). While there are different elements in this cement mix, we are primarily concerned with one ingredient, the IMF as a control factor and the Latin American debt as a symptom of diminished sovereignty.

THE INTERNATIONAL MONETARY FUND

Turning forty is thought to be a traumatic experience for most individuals; a landmark possibly signaling the onset of a midlife crisis. In 1984 the IMF and the World Bank celebrated their fortieth birthday with a colorful publication proudly proclaiming "Bretton Woods at Forty, 1944-84" and a large number 40 emblazoned at the center. Unlike ordinary mortals, IMF seemed to be suffering no midlife crisis, though interestingly enough, the world's international finances were in crisis in 1984 and the IMF itself is undergoing quiet but radical metamorphosis.

The IMF is poorly understood in the United States. According to John Eisendrath, Americans have never developed much interest in the institution because "the U.S. has never had to chafe under an austerity program imposed on it by the IMF." Consequently, the U.S. Congress does not exercise much oversight over an agency which dispenses 83 percent of U.S. multilateral aid, and the staff members of the U.S. House of Representatives committee with jurisdiction in this matter feel that hardly anyone in Congress has a grasp of the full details of the operations of the IMF (Eisendrath in Kojm, 1984: 107-116). The IMF itself puts out voluminous printed materials on its history and operations. However, more than one critic notes that these in-house publications are esoteric and seem to use language in a manner likely to confuse the layman.

The IMF was created in reaction to the problems that emerged in the period between the two world wars, when many nations engaged in protectionism and international trade declined to everyone's disadvantage. The United Nations Monetary and Financial Conference held at Bretton Woods, New Hampshire during 1-22 July 1944, was essentially dominated by Europeans and, more markedly, by American interests and concerns. The Conference considered the establishment of a sort of global Federal Reserve Board, with authority to regulate the global money supply and all other aspects of the world's monetary system. However, it backed off from such a global board in deference to national sovereignty, instead the IMF was created with authority to fix a quota for each member nation. This quota was based on the economic strength of each member, calculated in a manner in which was clearly intended to maximize the rule of the U.S.

Currently, the quota represented a sum of currency which each member country would give the IMF: 25 percent of the quota from each country is supposed to be provided in gold. The U.S. provided about one fifth of the Fund's resources, and any member nation has the ability to purchase, with its own currency, the currency of other countries to cover presumably temporary balance of payment deficits. Most of the Fund's literature and most literature about the Fund deals with this issue as if all Fund members were in some meaningful way equal partners in the arrangement. In fact, most countries buy dollars with their currencies. In the case of developed economies, they do so to correct the temporary imbalances mentioned earlier. Latin America republics simply buy up as many dollars as they can to cover what have become chronic balance of payment deficits.

How much can a Latin American republic buy? The answer can be very complicated; however, it should be noted that countries can take several times their quota, provided they tap into the various "facilities" or sub-funds within the IMF. For instance, in 1983, Brazil's quota was $1.1 billion. That year, Brazil arranged a credit line through 1985 for $6 billion. One loan from the Extended Loan Facility was worth $4.5 billion and represented 450 percent of this quota.

IMF members can use up to 25 percent of their quota with no questions asked. That 25 percent represents the first "tranche". Should they need additional money, they immediately begin to use subsequent tranches, and each succeeding tranche carries with it more requirements in the form of IMF conditions. For instance, a country wishing to use up to 100 percent or more of its quota will be required to submit a letter of intent explaining what steps it is taking to correct the situation which led to deficits and hence the need for the financial backing. The state of affairs wherein a country is held to specific commitments is called "conditionality", and the agreement which parcels out monies as conditions are met over a period of months or years is called a "stand by agreement". Even though the literature treats this whole area of conditionality as a sort of universal phenomenon, the fact is that conditionality is a one way street: developed countries tell third world countries how to set up their economic houses in order to receive whatever assistance they need from the IMF. While letters of intent and conditionality are secret, enough is known that we may simply put forth typical IMF conditionality requirements.

In nearly every case in which a Third World country is receiving assistance, conditionality involves all or some of the following prescriptions a recipient country must follow to receive the finding:

1. Reduce budget deficits and inflation by:
   a. diminishing and eventually eliminating subsidies to food and/or fuel;
   b. restricting and where feasible eliminating, future wage increases;
   c. reducing other social programs which increase the budget;
   d. tightening the money supply through higher interest rates and/or higher reserve requirements.
2. Reduce and progressively eliminate all exchange restrictions so as to ease trade and encourage foreign investment.
3. Encourage exports by devaluating currency.

Naturally governments could reduce deficits by increasing taxation and could improve their balance of payments situation by restricting imports and controlling hard currency overflows. They could declare a moratorium on debt repayment, from a debtors' cartel to bargain collectively for better credit terms or take any other combination of steps which might serve their needs. However, the IMF embodies the ideological views of Western Europe, Japan, and, most strongly, the United States. Its prescriptions simply oblige Latin America republics to open up their economies to corporate multinational investments; to free markets; to minimal governmental involvement, both in the economy and in the distribution of wealth within these societies in favor of the upper income groups. When subsidies to tortillas are removed in Mexico; when the price of fuel and transportation is raised, when special programs are eliminated; the lowest social groups are hurt the most. That is why IMF prescribed policies have generated rioting or civil unrest in the Dominican Republic, Peru, Brazil and Bolivia. Twenty percent of the population in Latin America lives on 2.5 percent of the national income and elimination of subsidies to food and fuel tends to push them over the edge to severe economic deprivation and near starvation.

Because of the monumental debt they carry and their consequent need for IMF assistance, Latin American government has lost most policy options. In fact, as matters stand in mid-1980s, these governments appear to have only one option---the implementation of IMF prescribed austerity. One unintended benefit of this situation is the regional return to democratic and constitutional governments. This return essentially results from the unwillingness of military leaders in the region to front for IMF and to carry out painful austerity measures. The military juntas, having generated much of the regional debt, are now more than happy to hand over the responsibility for implementing IMF austerity to hapless civilian leaders who, as with Raul Alfonsin in Argentina, must now toe the IMF line to make up for the debt accumulated during the 1970s when the Argentine junta and the international banking community had a cozy and easy relationship.

Politics is often defined as "the authoritative allocation of values" (Easton 1971: 136-137) or less esoterically, as the process whereby we determine "who gets what, when, how" (Laswell, 1936). Since at least the time of Aristotle, politics has also included the notion of sovereignty (Aristotle, 1962: 110) or ultimate authority. If we define national sovereignty as the ability of a state or government to make ultimate decisions regarding the allocation of resources within a given territory for a particular population, it is obvious that most Latin American republics have lost their ability to allocate resources independently. In fact, they now allocate resources according to IMF guidelines and sovereignty has been seriously impaired. Is this an aberration of the 1980s or simply evidence of a structural dependency of Latin America?

The monumental debt Latin America has accumulated is commonly explained as resulting from the circumstances peculiar to the 1970s. OPEC raised petroleum prices and generated large surpluses of cash which were subsequently invested in the U.S. and European banks. Latin American nations needed cash to pay for more expensive fuel, hence they borrowed lots of money from Western banks.

This stock explanation ignores the fact that all Latin American nations have the same debt problem. Brazil is a net importer of oil and it is a high debtor, but Argentina is self-sufficient in energy and has the same debt problem. Moreover, Mexico is a major exporter of oil and Venezuela is an OPEC member, and they have similar debt problems.

The debt problem can be partially explained in terms of circumstances of the 1970s. However, a more fundamental reason suggest itself to a political scientist with no particular claim to expertise in the complicated world of international finance. That reason simply put is the economic and political preponderance of the United States, the relative and ever increasing vulnerability of Latin American governments in their relations with the developed world, and the manner in which the world of international trade has been structured in the Hemisphere.

LAYMAN'S VIEW OF INTERNATIONAL FINANCE

Plato once postulated, through the words of Thrasymachus, that someone who was caught taking another's property was considered a criminal and a thief; if someone could do it a sufficiently large scale, as with a government taxing, then "you will hear no more of those ugly names..." (Plato, 1957: 26). People who print money in small quantities; are counterfeiters and their activities merit a special police force. Those who print money on a large scale-treasuries-are much honored; and if they print really huge sums they are not just a treasury, they are the U.S. Treasury.

Any government's treasury can print currency. We know that ancient goldsmiths began the practice of safekeeping their client's gold and issuing notes certifying the existence of the gold. In time people traded the paper certificates as if they were trading gold and it was not long before some enterprising goldsmith figured that he could issue more certificates than he had gold because of the small risk that all the gold in his care would be required simultaneously. We also know that ancient rulers engaged in a similar practice by issuing coins whose face value was greater than the metal content would satisfy. The difference between the face value and the actual value of the melted down metal plus the cost of coining was called "seigniorage". All modern governments fund themselves, at least in part through seigniorage; the U.S. reportedly has been deriving approximately $10 billion a year from seigniorage through the early 1980s and experts report that this activity is rarely discussed because "the creation of money is part of the central government activity..." (Grubel, 1984: 160). More likely, it is not discussed because we find it awkward to confess that a dollar is nothing more that a claim check against U.S. production of goods and services <u>created</u> by the U.S. Treasury. It is not a certificate for well protected gold stored in Fort Knox and the quantity of dollars is not given by God to the priestly brotherhood of the Federal Reserve Board.

John Locke had the right idea when he referred to money as something which "by mutual consent, men would take in exchange for the truly useful but perishable supports of life" (John Locke, 1952: 28). It could have been shells or sparkling pebbles, but by common consent we settled on "money". When our first Treasurer, Alexander Hamilton, urged in favor of the U.S. government taking on the national debt, he used an argument which showed how well the Founding Fathers and Framers of the Constitution understood money.

Hamilton said that if the government backed the debt, the paper then representing the debt would serve the purpose of money, and since the young republic was desperately short of cash, what better way could be found for killing two birds with one stone---resolving the debt problem and injecting a substantial sum of money into the economy. Said Hamilton:

> It is a well known fact, that in countries in which the national debt is properly funded, and an object of established confidence, it answers most of the purposes of money (Alexander Hamilton in Mason, 1965: 336).

Money in our times is nothing more than paper exchangeable for goods and/or services. The paper is exchangeable, not because of common consent (a quaint Lockean concept applicable only to the first generation of money users) but because its exchangeability is enforced by the state. Note the fine points on the ordinary one dollar bill---"this note is legal tender for all debts, public and private".

For most countries, exchangeability can only be enforced within the boundaries of the nation. Among nations, exchangeability is affected by a variety of factors. The U.S. dollar is more valuable than the Guatemalan Quetzal because it represents a claim check against a very large and varied economy. Dollars can buy computers, aircraft, surgical equipment, soybeans, etc. The selection of goods available from Guatemala is far more limited. But beyond that, the U.S. is better able to enforce the exchangeability of its currency.

The preponderance of the U.S. in international finance dates back to the end of World War II when the dollar was displacing the British pound as the World's currency. That preeminence was embodied in the Articles of Agreement of the International Monetary Fund adopted on July 22, 1944. Article III, Section 3 stipulates that 25 percent of each member country quota shall pay in gold the smaller of:

> (i) twenty-five percent of its quota; or (ii) ten percent of its net official holdings of gold <u>and United States dollars</u>... (emphasis supplied).

It is clear from this and other passages of the Articles of Agreement that gold <u>and United States dollars</u> are somehow the same.

Indeed, in 1944, and up until 1971, the U.S. dollar was convertible to gold at the rate of $35 per ounce. Even if everyone knew that like the ancient goldsmiths, the U.S. Treasury had printed a lot more dollars than it had gold, everyone understood the utility of the fictional convertibility between dollars and gold and the myth was preserved by nearly all foreign governments abstaining from converting dollars to gold. Of course, Charles deGaulle's France did not like to preserve a myth that appeared to benefit the U.S., and it exchanged the dollars for gold with seizing disregard for international financial conventions of the day. They did it so much that in 1971, the Nixon administration formally dismantled the dollar/gold myth.

The de-coupling of dollars from gold created new problem for the IMF which had been formed, in part, to police the convertibility of all nations' currencies into gold/dollars. What is important to understand here is that it was one thing for the U.S. to have its seigniorage, and an entirely different one for some "small time" Latin American republic to overheat the printing presses churning out currency. The job of the IMF was to maintain parity; no country could alter the value of its currency by more than 10 percent without IMF clearance.

With the dollar no longer pegged to gold, one would have imagined the IMF quickly becoming obsolete. But, the IMF had been developing a new lease on life since the 1960s by creating a gold substitute---Special Drawing Rights, SDRs for short. Originally, SDRs were to have a dollar par value. Up to August 31, 1974, one SDR was equal to one U.S. dollar. From that time on the value of one SDR was essentially indexed to a large number of "hard" currencies. Beginning on January 1, 1981, the value of an SDR has been determined by taking into account the value of five currencies in the following proportions: U.S. dollar, 42 percent; German mark, 19 percent; French franc, Japanese yen and British pound, 13 percent each. Here again, the preponderance of the U.S. dollar is continued by virtue of its unique role in determining the SDR values.

This preponderant role of the dollar in determining the value of SDRs enables the U.S. to operate by different rules. While applauding and encouraging the IMF to dictate austerity for Latin America, the U.S. government engages in large scale deficit spending. From 1983 to 1984 alone, the U.S. doubled its balance of trade deficit to over $100 billion. In an article published in 1984, <u>The Wall Street Journal</u> speculated as to what it would be like if the IMF were to dictate to the U.S. the same sorts of austerity measures it regularly imposes on Third World debtors. After musing about the specific cuts and the pain they would cause, the authors went on to reassure the readers that no such conditionality would ever apply to the U.S. because the U.S. debt is not the result of borrowing other countries' currencies. The article concluded by candidly declaring that "when the U.S. government wants to pay its debts, it can always print dollars, whatever the inflation consequences; other countries don't have that luxury" ("Just Imagine an IMF plan for America", <u>The Wall Street Journal</u>, 22 June 1984: 26). In fact the record of 1984 and 1985 seems to support the proposition that thus U.S. can also finance its debt by maintaining high interest rates and this attracting new Eurodollars into the U.S. with minimal or no impact on inflation.

Harry Magdoff quotes an Exxon senior economist to the effect that "...after a hundred years of international monetary conferences, men still have not resolved their differences. The answer lies in one word----power". Indeed, this same individual felt that Bretton Woods was simply a struggle for power and Magdoff goes on to draw his own conclusions:

> It should be obvious that a key currency country obtains clear-cut advantages. It can operate in the arena of international commerce and finance with a much greater degree of freedom than a run-of-the-mill nation, let alone a Third World country. If it overbuys from other countries, it can pay for these purchases by merely printing or otherwise creating more domestic money. It therefore, can live for long periods quite comfortably with balance-of-payments deficits; instead of having to tighten its belt, it can become wealthier from an excess of imports over exports. Even more important, the ability to create unilaterally additional international money expands a country's ability to export capital, thereby enabling to export capital, thereby enabling it to obtain assets yielding a steady return flow of interest and dividends (Harry Magdoff in John Stack, 1983:94-95).

THE LATIN AMERICAN DEBT

As <u>The Wall Street Journal</u> so well understands, Latin American republics cannot pay their debts by printing more currency. They owe dollars to U.S. and European government and private banks. Their debts have mounted over the years to staggering proportions. By 1984 Argentina owed approximately $45 billion; Brazil $95 billion and Mexico $90 billion. Because interest rates have increased on outstanding loans, the amount of income required to service the foreign debt has increased to the point where all three countries would require more money to service their debt than they have income from the export of goods and services.

Most analysts prefer to explain the Latin American debt problems as the result of unfortunate and almost accidental circumstances outlined earlier in this paper. I propose to explain the Latin American debt problem and the resulting loss of sovereignty as the logical and inevitable outcome of Latin American power deflation <u>vis à vis</u> the U.S., Western Europe and Japan. The IMF is the enforcement mechanism entrusted with the task of maintaining this status quo. Naturally, the IMF views matters differently. Its standard explanation of current affairs runs as follows:

> The large changes in exchange rates that have taken place among the major currencies over the past years have had a substantial impact on the developing countries, not only on their effective exchange rates ... but also on such variables as commodity prices, the real value of international reserves, and external debt. In particular, the dollar appreciation tended to depress commodity prices in the U.S. dollar terms while nominal magnitudes of debt service (which was mainly in dollars) remained unchanged, thereby raising the real burden of debt service... (IMF, 1984: 49).

In this tense bureaucratic language, the IMF presents the plight of Third World countries as a circumstance emanating from what the insurance business would call "an act of God". There are droughts and hurricanes, and "large changes in exchange rates" which, from time to time impair the welfare of a nation or groups of nations. Indeed, the evidence suggests that Latin America's economic performance has suffered greatly. For instance, form 1960 to 1973, Latin American economies had an average annual growth of 5.6 percent compared with 4.9 percent for the industrialized market economies. From 1973 to 1979, again Latin America grew at an average annual rate of 5.0 percent versus 2.8 percent from the western developed countries. In 1980 the Latin American economies grew by 5.8 percent while the industrialized market economies grew by a meager 1.3 percent. From that year, and with the exception of 1982, we see the industrial market economies generally outperforming the Latin American economies, so that by 1983 the Latin American economies shrank by 2.2 percent while the industrial market economies grew by 2.3 percent (The World Bank, 1984: 11). For Latin America, the Gross Domestic Product (GDP) fell by almost 6 percent in 1983. By 1983, the region's debt was estimated at $336 billion (it had been $75.4 billion in 1975) and that debt represented the equivalent of 56 percent of the regional GDP and 325 percent of the region's annual exports of goods and services (Jorge Espinosa-Carranza, 1984: 8). According to the World Bank, by 1984 the region's per capita output had fallen to approximately what it was in 1976. By mid-1984, Latin America was facing the worst financial crisis since the Great Depression and experiencing economic declines of 3-4 percent on a yearly basis.

While the IMF, the World Bank and others generally describe the plight of Latin American economies as resulting from acts of God; they also have a way of suggesting that to the extent that anyone can be faulted for the problem, it is the Latin Americans themselves. There is a sort of unanimity of this issue among the U.S. government, the World Bank and, of course, the IMF. For instance, speaking in Buenos Aires in late 1984, the U.S. Ambassador to the Organization of American States (OAS), J. William Middendorf II, blamed the high rates of inflation in Latin America on "undue monetary growth" and more particularly on the "relatively high levels of government ownership of enterprises". In fact, it is the traditional involvement of Latin American states in the economy and the development of state run enterprises that seems to offend, most deeply, the fundamentalist free market faith so strong in Washington in the mid-1980s. After describing the calamities of Latin America, Ambassador Middendorf said:

> These circumstances are exacerbated by the official bias in favor of the public sector and a pattern of increasing government encroachment. It is now becoming clear that private enterprise in Latin America is in a fight for survival. (U.S. Department of State, Current Policy No. 638).

The U.S. Ambassador's specific charges are that the Latin American governments have increased ownership of enterprises, expanded regulation of private economic activity, increased government use of GNP and government investment in economic enterprises. To correct the situation, the

Ambassador suggested a number of steps in addition to doing away with the Calvo Doctrine (that foreign countries should submit to the laws of the host country), Latin American countries should have above all, "unrestricted transfer of capital, compensation, and other payments into and out of the host country".

This prescription well represents the views of the current U.S. President, Ronald Reagan, who states that "millions of individuals making their own decision in the marketplace will always allocate resources better than any centralized government planning process". (U.S. Department of State, Current Policy No. 513). This simple and most fundamental faith of the President also shows in the statement of the U.S. International Development Agency (AID) before the Congress. AID has gone on record saying that, U.S. aid is guided by "development through private initiative...". The Agency has also pledged itself to encourage policy reform in the Third World so as to "foster a free and open climate in trade, private financial flows and the LDS' domestic market". (U.S. Agency for International Development, 1985: 39).

Given their common origin at Bretton Woods, it is not surprising that the World Bank and IMF have very similar views, and that those views in turn, closely match those of President Reagan and AID. In fact, IMF conditionally represents the imposition of U.S. views on the internal affairs of the sixteen Latin American states which, since 1981, have sought emergency balance-of-payments support from the IMF. Ambassador Middendorf, President Reagan, the World Bank, IMF and AID are unanimous in the belief that a free market economy and minimal state intervention in the economy maximizes individual freedom, and can lead to more economic development. The World Bank and official spokespersons for the U.S. government are particularly fond of noting that while Latin America has been lagging behind, nations in Asia and the Pacific (South Korea is often mentioned) have prospered by attracting larger shares of U.S. overseas investments. U.S. Secretary of State, George P. Schultz, addressed the OAS in November of 1984 saying that "attracting both domestic and foreign investment can be a route to _more_ freedom and independence rather than less". (U.S. Department of States, Current Policy No. 638). In the face of such unanimity regarding the historical circumstances that led to the high level of debt and the cures for the problems facing Latin America, what alternatives do Latin American governments have?

As a short term solution, Latin American debtors could form a cartel and through collective bargaining secure a refinancing of their debt in such a fashion that debt service would be kept to a reasonable level---say 25 percent of annual income from exports of goods and services. Failing that, Latin American governments could simply default on their debts; something which Bolivia has essentially done and Costa Rica has partially done. In addition to these short term solutions, Latin American republics could opt for long term protectionism to shield their economies from the unlimited competition of world markets.

Citing a Brookings Institute study, The Wall Street Journal speculated that Brazil might be better off by 1987 by repudiating its debts and that Argentina would probably survive with relative ease. On the other hand, Colombia, Peru and Chile would fare far worse. (The Wall Street Journal, 22 June 1984: 27). In addition to the obvious size of their economies, one of the reasons why Argentina and Brazil could possibly endure default and the subsequent reprisals, is that for years they ignored the doctrines of free international trade and protected an emergent national industrial base. The point that should be emphasized here is the "delinking", as the World Bank calls the policy of substantial or limited economic protectionism, can and does work.

A completely free international trading system with no exchange restrictions, "encourages optimal use of available investment resources," to quote from the World Bank's 1984 report. However, optimal use in a global sense simply means that a country like Argentina would give up all its "inefficient" manufacturing and revert to an agrarian economy dominated by beef exports. After all, Argentina has a natural or comparative advantage with a few other products, but Argentine auto, ship building and other industries cannot compete with their counterparts in the industrial or developed market economies. It may be inefficient in the eyes of the international financial community for Argentina to have an auto industry; but if the Argentines take pride in producing their own cars and want to pay the higher prices for less efficiently produced domestic cars, it is not irrational for the government to adopt such policies as will preserve an auto industry. Through Fabricaciones Militares, a government owned conglomerate of defense oriented industries, the Argentines have produced a variety of goods essential to their national security for decades. When they entered into war with Britain, they found that they could count on their domestic defense industry. They also found that the only country which would supply them with computers during the conflict was neighboring Brazil. Brazil had also amused the international business community by insisting on a domestic computer industry regardless of how expensive these domestic computers would be.

In decisions regarding the economic policy, allocation of resources is a matter of choice that sovereign governments should make. There is no universally accepted sacred writ which prescribes integration into the world economy as the only rational policy alternative. There is certainly no reason why the Western powers should utilize the vehicle of the IMF to limit Latin American governments to one policy option, however sincere these powers might be regarding their faith in the universal benefits of a fully integrated world economy. As an example, India successfully delinked its economy and still has managed a very steady 3.5 to 4 percent rate of annual growth over three decades.

President Reagan has called on the IMF, the "linchpin of the international financial community" to coax the world economy to grow by playing "the Dutch uncle,' talking frankly, telling those governments things we need to hear but would rather not" (U.S. Department of State, Current Policy No. 513). Naturally, "those governments" Reagan refers to, have no choice but to pay attention. The IMF can give or take away; it can even coax private bankers in the U.S. and Europe to roll over all loans or come up with new ones, provided "those governments" come up with acceptable letters of intent and then adhere to the terms of conditionality. In the IMF only six countries appoint directors—the U.S., Britain, Germany, France, Japan and Saudi Arabia. The U.S. alone controls 19.32 percent of the votes; the next largest vote is that of Britain, 6.7 percent. The elected directors are chosen by groupings of countries which are gerrymandered to maximize the influence of Western Europe, the U.S., Canada, Japan, Australia, and New Zealand. Thus, for instance, most of the small Caribbean states are grouped with Canada, and thus are represented by a Canadian; Cyprus, Rumania and Yugoslavia are together with the Netherlands. The result of this proportional or weighted voting is that the U.S. alone controls nearly one-fifth of the vote; the U.S. and major European powers and Japan (key-currencies countries) control 41.28 percent of the vote. If to that we add other industrial market economy directors representing groupings which might include Third World countries, we have an overwhelming 65.25 percent of the IMF votes. Latin Americans get 10.27 percent of the vote and presumably could count on some votes in the remaining 24.48 percent of the vote controlled by the rest of the Third World; however, this 24.48 percent includes many votes controlled by middle and high income oil exporting countries whose interests on financial matters tend to coincide with those of the U.S.

Given the close relationship between the U.S. and the IMF, it is not surprising that relatively few Latin American countries have dared to default on their debts. In 1961, Cuba defaulted with well known historical consequences. More recently Bolivia has virtually defaulted by temporarily suspending interest payments. However, Bolivia is considered an international basket defaulting out of sheer economic and political chaos, rather than as a result of a policy decision.

The only Latin American republic to actually take on the IMF and the international banking community in recent years is Argentina. This Southern Cone country has a number of qualities which render its position substantially stronger than is generally the case in the Third World. Argentina is virtually self-sufficient in energy and is among the few countries in the world producing a net surplus of food. In short, if the international financial community were to ostracize it, Argentine transportation would continue to run and people would be fed - something that relatively few countries in the world can claim.

During the first quarter of 1984, Argentina began to play a curious game of brinkmanship. It declined to pay interest on outstanding loans while negotiating dangerously close to the deadlines by which U.S. bankers would have to classify their Argentine loans as "non-performing" and thus face the prospect of reporting considerably low earnings to Wall Street and their stock holders. These deadlines are particularly troubling to U.S. banks which operate under Securities and Exchange Commission rules more stringent than those covering European banks. Reported lower earning can lead to runs on deposit by nervous bank clients and possibly the consequent failure of some of the U.S.'s largest banks. As the deadline approached with the end of March, 1984, U.S. Secretary of Treasury, Donald Regan appeared to cave in. On Tuesday, 27 March 1984, with just a few days remaining, he publically suggested that U.S. regulations requiring the reporting of lower earnings should be relaxed. The suggestion created such excitement that on Wednesday the Secretary was quoted as retreating from this position and saying that "we are not going to bend the rules". Indeed these rules were not bent, and a deal was wrote out at the 11th hour- Mexico, Venezuela, Brazil, Colombia and eleven private banks came up with $400 million so that Argentina would, with $100 million of its own reserves meet the deadline. What was significant was that the eleven private banks provided Argentina with $100 million at one eight of a percentage point over the London Interbank Offered Rate (LIBOR). The normal rate would have been less than one and one half points over LIBOR. In return for this clearly concessionary loan from the private banks, Argentina stalled trying to force the IMF staff to come up with a letter of intent which would preserve the political head of newly elected President Raul Alfonsin. President Alfonsin attempted a new approach; he circumvented that IMF staff and tried to negotiate with the Directors. Alfonsin was rebuffed, and during 1984, the tug of war between Argentina, and the IMF with the support of the international banking community, continued. In time, the banks began to declare parts of their Argentine loans "non-performing," as a means of playing their own game of brinkmanship. If the banks could phase in their Argentine losses over a period of several quarters, there would be no run, or so they hoped. In time, Argentina came to terms with the IMF, but not before some very tough negotiations. According to Fortune, the IMF and the banks described the painful consequences to Argentina, should it default. The U.S. Treasury came up with a list of items that would become scarce in countries defying the IMF. R. T. McNamara, Deputy Treasury Secretary said that the list raised such questions as: "Have you ever contemplated what would happen to the president of a country if the government couldn't get the insulin for its diabetics?" (Gary Hector, 1985: 36). President Alfonsin was certainly familiar with what had happened to his predecessor, General Leopoldo Galtieri after he lost the war with Britain over the Malvinas, and he ultimately lacked the nerve to defy conditionality.

Again, during the Spring of 1985, Argentina began its annual round of flirtation with default. The interaction between this major South American nation, the U.S. and the international lending agencies during the first half of 1985 provides an excellent illustration of the erosion of governmental power in Argentina and a corresponding assumption of power by the IMF and through it, the U.S. government. The implication here is not that the Argentine government lost all power; indeed that government has partially frustrated the demands for "structural reforms" emanating from the Reagan administration.

After the unsuccessful attempts to circumvent the IMF staffs in 1984, President Alfonsin replaced Economics Minister Bernardo Grinspun with a man described as a technocrat who would work more easily with the IMF (The Wall

Street Journal, 20 February 1985: 33). By the time Sorrouille took over in February, 1985, the Argentine debt had grown to $50.4 billion and the new minister asked the IMF for the time to develop a plan of action. During March, President Alfonsin visited Washington seeking financial assistance from the U.S. and promoting acceptance of the fact that Argentina did not have much of a chance of meeting IMF inflationary targets (150 percent for 1985, down from 688 percent in 1984). His visits followed the February cancellation of a 15 month stand-by agreement worth $1.7 billion with the IMF. This cancellation had come at a time when the country was behind in its payments to commercial banks by some $800 million, and in fact President Alfonsin seems to have been seeking a $500 million bridge loan to tide his government over until some new deals could be worked out with the IMF. As in 1984, Argentina was again on the brink of defaulting at the end of the first quarter of 1985.

President Alfonsin was cordially received by President Reagan, Congress and the National Press Club, but he garnered little comfort. In fact, during the months following his visit, the Reagan administration went into its annual spring rampage giving sermonets on the virtues of governmental disengagement form the economy and lower taxes as the panacea to all problems. In a speech given in Madrid on May 7th, President Reagan again equated individual freedom with a minimum rule for the state in the economy and credited implementations of this ideology with creating high standards of living in the West. Without mentioning any governments in particular, the U.S. President denounce subsidized exports, protectionist tariffs and governmental controls. He also explained that "...in Asia, economic freedom has really taken hold, fueling the meteoric rise of the Pacific Basin nations..." He even credited economic freedom as "...even giving communist China a helpful push towards prosperity," thus implying that communism is compatible with economic freedom and its equivalent, individual freedom, so long as it opens markets to international business. On May 20th, Elinor G. Constable, Acting Assistant Secretary for Economic and Business Affairs, delivered a speech in Los Angeles attributing U.S. prosperity to lower taxes and governmental disengagement from the economy and calling on other countries to "improve their investment climates," (U.S. Department of States, Current Policy No. 708, Washington, May 1985).

On May 21, David C. Mulford, Assistant Secretary of the Treasury for International Affairs warned Latin American nations that the capital needs of the area would not be met by means of more U.S. aid, nor with more exports from that region to the U.S. market. He called for "fundamental changes in the region in order to avoid a new debt crisis (Clyde H. Farnsworth, 1985: 6). The word "fundamental" is significant here because it has emerged as a key to the differentiation between the typical conditionality terms offered by Latin American countries to the IMF, and the kinds of governmental disengagement prescriptions so persistently proposed by the Reagan Administration.

Governmental disengagement is not what the international commercial banking community advocates when faced with potential losses. In May, Argentina's third largest bank, Banco de Italia y Rio de la Plata collapsed, and U.S. bankers demanded that the Argentine government assume responsibility for the $200 million debt owed them by the failed banking institution as a condition for their continued participation in IMF stand-by agreements. In June, the Alfonsin Adminstration bowed to pressure and assumed the debt. This development well illustrates the ambivalent signals that Latin American debtors receive. The U.S. and the IMF want the Latin American governments to institute a 19th Century version of economic liberalism, except where that policy might harm U.S. interests.

Also by early June, Argentina was again in desperate need for help. The government was facing the prospect of having its debts downgraded by American bankers in compliance with U.S. regulations because the country was nearly six months behind in interest payments of $1.2 billion. Again, the IMF would have to agree to a new plan so the another "bridge" loan could be arranged. The U.S. regulators were considering downgrading Argentine loans by one notch to "substandard" or by two notches, to "value impaired". The fist step would cause smaller banks to pull out of future financing, the second step would require all banks involved to set aside funds against their Argentine liabilities, thereby reducing current profit margins. It should be noted here that U.S. banks seem comfortable with the notion that U.S. regulations dictate how the crisis must be managed, and in this area, the Reagan Administration seems disinclined to practice its regulations regarding this area than most Western European governments. At the last minute, the IMF and Argentina once again pulled off a face-saving agreement which avoided default. The U.S. and presumably other Latin American government loaned $400-500 million to partially cover overdue interests and satisfy U.S. banking regulators.

The interplay of the U.S. Treasury, banking regulators, the banks, the IMF and Argentina has a certain comical ring. The IMF dictates the conditions which Argentina must accept so that Treasury will put up the money, so that Argentina can pay the interests, so that the banks can satisfy the U.S. regulators. A by-product of this relationship is that Argentina, a theoretically sovereign nation, functions as a virtual dependent of the U.S., caught between different levels of the Washington bureaucracy, and trying to please all.

By June, 1985, Argentina paid $250 million and thus was current in its interest payments to the end of 1984. Argentina attempted to keep its overdue payments no more than approximately six months behind schedule. However, it was clear that the U.S. and the commercial banks were continuing their relentless pressure for some "fundamental" reforms, and that in fact, "bridge" loans would not be forthcoming unless such reforms were adopted. The Alfonsin government eventually had to agree to devalue the peso by 15 percent; to slash spending; to improve severe controls on money and credit; to freeze all government hiring, and to reduce subsidies (and hence increase prices) of gasoline and electricity, to raise fees for government services, and, most significantly, to limit wage increases to 90 percent of inflation. Among the key negotiations of the agreement was David C. Mulford, the Assistant Secretary of Treasury who had lectured Latin American nations earlier in May regarding the need for "fundamental" changes.

On June 14th, President Alfonsin announced the creation of a new currency and pledged his government not to print more money that it took in. The new currency, called the Austral, essentially meant the logging off of three zeros from the old currency---an operation with which Argentines have had previous experience. The same day that these most severe steps were announced, the U.S. banking regulators stated that they would not downgrade the Argentine debt. As a fact saving device for President Alfonsin, the Treasury Department's Mulford declared that President Alfonsin had gone beyond the requirements of the IMF. (Lydia Chavez, 1985: 1)

While these events are subject to interpretation, it is reasonable to assume that President Alfonsin was pressured into adopting internal political measures which he found difficult, if not dangerous to implement. Given the traditional power of the Peronist Labor movement it is also clear that limiting wage increases the 90 percent of inflation is the equivalent of decreasing a 10 percent annual wage cut for Argentine workers. If politics is the determination of who gets what, when, how (as Harold Laswell contends), then it is clear the Argentine workers get a 10 percent wage cut, every year the IMF stand-by agreement is in effect, by governmental policy; it is also clear that they and other Argentines get to pay more for gasoline and utilities, and for the diminished government services they might receive.

Did President Alfonsin please the international banking community? To the extent that The Wall Street Journal reflects those views, we might note that in an editorial for the 28 June 1985 issue, The Journal gave President Alfonsin "high marks for bravery," but then went on to express doubts about his success because he had not really implemented "structural" reforms; for example, "dismantling state enterprises". As with the Reagan Administration's "fundamental" reforms, The Journal's "structural" reforms means "opening the chance for markets to work," as the nations of the Pacific rim have done. It should be noted that both The Journal and The New York Times carried personal interest stories during 1985 graphically depicting the plight of individual Argentine families, one of the victims of political violence, another the victim of economic hardship. Only the Times attempted to connect the hardships of a working class Buenos Aires family with IMF agreements (Lydia Chavez, 1985: E5).

Writing in the mid-1970s, Cheryl Payer described debt peonage; a situation in which the worker is unable to use this nominal freedom to leave his employer because the employer provides him with credit necessary to supplement meager earnings for purpose of survival. She added:

> Precisely the same system operates on the international level. Nominally independent countries find that their debts and their constant inability to finance current needs out of imports, keep them tied by a tight leash to their creditors. The IMF orders them, in effect, to continue laboring on the plantation, while it refuses to finance their efforts to set up in business for themselves (Cheryl Payer, 1974: 49).

Indeed, this analogy seems closer to the truth than the alternative often implied by the international financial community---in which careless, profligate nations spend others' money in irresponsible schemes. In President Reagan's imagery, the spendthrift nations are in need of a trip to the woodshed with their no nonsense, thought "Dutch uncle," the IMF. The options open to Latin American nations are not as few as one might imagine. If their situation were to become sufficiently desperate they could take a variety of approaches. For instance, they could encourage the swap of debt, for equity. Chile has been encouraging potential investors to buy its outstanding, dollar denominated debt at discounted prices. The Chilean government then buys that obligation with Chilean currency, but at face value dollar-equivalent, and the investor in turn, buys Chilean assets with currency purchased. In effect, the investor can have his purchasing price reduced by the equivalent of the discount it gets from the financial institution selling the debt instrument. Since the government has been trying to lure foreign investors into the country since the overthrow of the Salvador Allende government, this new approach is merely a variation on an approach rejected by some Latin American nations. We should note that for many years, the policy of civilian governments called for "Chileanization" of capital assets and that the legitimacy of the new debt/equality sway may commensurate with that of the current military regime.

In countries where the debt problem is closely related to capital flight, governments could try to do what Mexico is reported to have started in July of 1986; to issue savings certificates with principal and interest redeemable at maturity in pesos, but then peso/dollar exchange. In effect, the Mexican government is issuing a kind of Mexican dollar, pegged to the value of the U.S. dollar. If these instruments could sell internationally, they would represent a potentially revolutionary way for Mexicans to cut into the printing monopoly of the U.S. Treasury.

Latin American republics with sizable dollar reserves could also leverage those reserves by issuing dollar denominated debt instruments; again, Latin dollars. If done moderately, Lain Americans could initially issue 110 percent of their dollar holdings, with the same expectation of ancient goldsmiths---that creditors would not all demand payment at once. They could do with their paper debt what the U.S. did with the gold in Fort Knox---turn into a seed for a credible new international currency.

Peru has attempted to deal with its debt problem by limiting its debt service problems to no more than 10 percent of its export income since 1985. As of 1986, Peru had paid a total of $350 million in lieu of the approximately $2 billion dollars due on this $13 billion foreign debt. The IMF itself received a $35 million instead of $190 million due on August 15, 1986 and it promptly declared Peru ineligible for any new loans. Thus Peru became the first Latin American republic to be ineligible for help and joined the ranks of Guyana, Liberia, the Sudan and Vietnam in the IMF's black list. While this approach is certainly innovative, its chances for success would be considerably enhanced if the policy were to be the result of collective action among the region's debtors.

Finally, the Latin American republics can, and in fact have attempted to form debtor's cartels for the purpose of engaging in collective bargaining with the IMF and the international financial community. This approach would be extremely advantageous and, after all, that is what creditors have done in dealing with Latin American republics. Why then, have Latin American republics failed to come together for debt bargaining purposes? A complete answer would rather be complex; however, it should be noted that the U.S. and international bankers have strongly opposed the move and that whenever a cartel was in the offing, the U.S., has been able to obtain the defection of at least one or more major debtors with some sort of concessionary treatment.

CONCLUSION

In his First Report on the Public Credit, on January 14, 1790, Alexander Hamilton said:

> Those who are most commonly creditors of a nation, are generally speaking, enlightened men; and there are signal examples to warrant a conclusion that, when a candid and fair appeal is made to them they will understand their true interest too well to refuse their concurrence in such modifications of their claims as any real necessity may demand.

In line with Hamilton's insight, we may hope, perhaps predict, that the IMF and the international financial community will soon realize that the Latin American debt problem is not a temporary aberations, but a symptom of an unequal system structured in such a fashion as to trap and definitely hold these debtor nations in a form of peonage called conditionality. To the extent that local populations come to terms with the power deflation of their national regimes, they are likely to take increasingly more desperate action. This prospect calls for significant modifications of the claims against debtor nations and of the system which has so burdened them. Surely the IMF and the U.S. leaders will understand the Latin American desperation over the debt issue and will act accordingly.

REFERENCES

Aristotle (1962). *The Politics*.
    New York: Penguin Books.

Easton, David (1971). *The Political System - An Inquiry Into the State of Political Science* (2nd ed.). Chicago: The University of Chicago Press.

Espinsoa-Carranza, Jorge (1984). "The External Debt: An Analysis and a Look into the Future." *IDB News-Monthly Newsletter of the Inter-American Develop Bank* (May).

Farnsworth, Clyde H. (22 May 1985). "US Urges Changes in Latin Politics" *New York Times*.

Hector, Gary (1985). "Third World Debt: The Bomb is Defused." *Fortune*. Vol III, No. 4 (February 18):36-50.

Grubel, Herbert G (1984). *The International Monetary System* (4th ed.) New York: Penguin Books.

International Monetary Fund (1967). *Articles of Agreement of the International Monetary Fund*. Washington: IMF.

_____ (1984). *Bretton Woods at Forty*.
    Washington: IMF.

_____ (1984). *Annual Report*.
    Washington: IMF.

Kojm, Christopher A., ed. (1984). *The Problem of International Debt*. New York: The H.H. Wilson Company.

Laswell, Harold (1936). *Politics: Who Gets What, When, and How*. New York: McGraw Hill.

Locke, John (1952). *The Second Treatise of Government*. New York: Bobbs Merrill.

Mason, Alfred T. (1965). <u>Free Government in the Making</u> (3rd ed.) New York: Oxford University Press.

Modelski, George, ed. (1979). <u>Transnational Corporations and World Order</u>. San Francisco: W.H. Freeman.

Payer, Cheryl (1974). <u>The Debt Trap – the IMF and the Third World</u>. New York: Monthly Review Press.

Plato (1957). <u>The Republic</u> (translation by A. D. Lindsay). New York: E.P. Dutton.

Stack, John (1983). <u>Policy Choices – Critical Issues in American Foreign Policy</u>. Gulford, Conn: Dushkin Publishing.

U.S. IDCA (1985). <u>Congressional Presentation Fiscal Year 1985</u>. Washington: U.S.I.D.C.A.

World Bank (1984). <u>World Development Report 1984</u>. New York: Oxford University Press.

10

Innovation in the IMF: The Cereal Imports Facility

Valerie J. Assetto

After the world food crisis of 1972-1974, scholars began to investigate food crises in terms of food security, the ability of governments to assure a minimum, stable level of consumption for their citizens. Following the collapse of talks designed to ensure food security by establishing an international buffer stock facility in grains, the concept of a food financing facility was introduced in which international financial assistance would be provided to countries experiencing a large short-term increase in the price of their food imports. Because of its pivotal role in the provision of international financial assistance, the International Monetary Fund soon became the international organization of choice for administering this facility. The IMF's Executive Directors approved the Cereal Imports Facility (CIF) in May 1981 (Decision No. 6860-(81/81); IMF, 1981).

This study analyzes the Cereal Imports Facility in the context of innovation in the IMF using primary and secondary sources. The origins, structure, and fate of the CIF are examined from 1981 through 1985 in order to evaluate the impact and utility of the CIF in ameliorating the food security problem. In a broader context, the CIF also illuminates the process of innovation in the IMF since 1981. After placing the CIF in the context of IMF attempts at innovation, a brief history of the origins of the CIF is provided, using secondary sources, to illuminate the original purposes and goals of the facility. The actual performance of the facility is then examined with respect to its utility for developing countries.

INNOVATION IN THE IMF

Scholars of public administration and organization theory contend that public organizations are slow to adopt innovations in their procedures.1 According to this argument, this reluctance to innovate stems both from an organization's internal structural characteristics and its environment. Typically, these organizations are publically-owned and accountable to a large constituency. Favorable reviews by powerful members in that constituency are essential to the organization's survival, yet the criteria used in the evaluation of the organization are often vague and may conflict with the criteria used by other members. Organizational mistakes, therefore, can be fatal, a decided disincentive to innovation.

Hierarchial decision structures and heavy reliance on standard operating procedures (SOPs) also inhibit change in organizations. Again, the costs of erring are increased under this type of structure so that any deviation from SOPs must demonstrate that it will result in a marked increase in benefits for the organization. These benefits are often measured in terms of the response of constituents of the organization or in its environment which control essential resources (such as the groups which control the organization's budget).

A public organization's environment also tends to reduce incentives to innovate. Because most public organizations operate in an oligarchic, or even monopolistic, environment, there is no pressure from competitors which would force the organization to adapt in order to survive. In an oligarchic environment, therefore, both the costs and benefits of innovation are reduced.

What prompts innovation, then, in public organizations? The survival imperative eventually forces adaptation, particularly in times of crisis. By definition, a crisis threatens the existence of an organization which must react to such significant changes in its environment in order to retain legitimacy. As is discussed above, the positive evaluation of powerful members of the organization and/or the environment is vital to that organization and failure to respond appropriately to a crisis would endanger the organization's reputation with that group. As a result, it is reasonable to expect that crises which directly affect the most powerful members of an organization will result in the greatest pressure to innovate. Conversely, crises which affect a relatively minor subgroup within the organization or its environment will evoke a less creative response. Standard operating procedures should prevail in this case.

The IMF can be characterized as a public organization, but it also possesses some of the attributes of a private organization, particularly with respect to ownership.[2] Nonetheless, the IMF is not noted for its innovative behavior, particularly with respect to the problems of developing countries. Like most public organizations, the IMF is slow to introduce new facilities whose potential political and economic costs might damage the legitimacy, and ultimately the survival, of the organization. Thus, innovations in financing facilities are most likely to be adopted by the Fund only at those times when the organization is threatened by events or when powerful actors in its environment desire the change. Non-systemic crises evoke responses in the IMF which tend to rely on standard procedures since the risks of innovation at these times are great. Therefore, during crises whose effects are limited to the Third World, the IMF will attempt to absorb new, Third World-oriented facilities within the Fund's established structures by extending old facilities or merging new procedures within established structures.

For example, the oil crisis of 1973-1977 stimulated the adoption of five facilities in the IMF designed specifically to benefit Third World countries. The oil crisis was a system-wide crisis which threatened both developed and developing countries alike. Developed countries such as Italy and Britain required substantial amounts of IMF assistance during this period in order to finance balance of payments problems stemming from the sharp increase in oil prices. As a result, the most influential actors both within the IMF and in its environment, the U.S. and the E.E.C. states, eventually supported the new facilities; the considerable amount of votes in the IMF controlled by these countries virtually guarantees failure for any proposal which does not meet their approval (the U.S. alone possesses over 19 percent of total IMF votes).

In contrast, food insecurity is basically a problem for developing countries which do not possess the financial resources required to meet sudden changes in the price and volume of their food stocks. The developed countries in general do not suffer from instability in their food stocks; many are major exporters of agricultural products. Food insecurity crises are thus Third World crises and according to the argument advanced above, attempts to mitigate this problem through innovation in IMF procedures will probably result in failure or alteration of the proposed innovation to conform to standard Fund practices. The history of the Cereal Imports Facility confirms this.

The Cereal Imports Facility (CIF) provides an excellent example of Fund innovative behavior in a non-system crisis. The CIF was originally proposed as a separate facility within the IMF and was novel not only in its intention but also in its origins. The impetus for the CIF primarily came from outside the organization through two other international organizations and the facility was intended to address a problem which almost exclusively affected developing countries. Thus, if the argument advanced above is correct, in this situation the IMF would not be compelled to innovate and would attempt to absorb this new idea into its standard operating procedures.

The CIF is also interesting in itself as it represents one of the early attempts to address the food security problem. The facility has now been operating over five years, during which Africa has experienced a severe food crisis. It is reasonable to expect that if the intentions of its founders have been met, the CIF should have played a significant role in alleviating the crises. This analysis investigates the extent to which the CIF has been a useful resource for developing countries suffering from food insecurity or merely a symbol of developed countries' concerns for the problem. Thus, the CIF is a good test of both the IMF's creativity in a non-systemic crisis and its responsiveness to the problem of developing countries.

THE CEREAL IMPORTS FACILITY

The world food crisis of 1972-1974 prompted the investigation of several alternatives designed to alleviate or eliminate the problem of food insecurity. The problem, as it was now defined, was not a lack of world food supplies but poor international distribution networks which impeded the flow of adequate food supplies to poorer countries which did not possess the financial reserves required to purchase food stocks during period of domestic food production shortfalls (Green and Kirkpatrick, 1981; Adams, 1983; Huddleston et al., 1984; Murdoch, 1980). While scholars urged the adoption of domestic policies which would stabilize domestic production of food, especially grains, there was a widespread agreement that rapid progress could be made toward reducing the role of the international trade and financial systems in the food security crisis through some type of international food arrangement which would stabilize the world's supply and price of foods generally imported by developing states (Green and Kirkpatrick, 1981; Adams, 1983; Huddleston et al., 1984).

As identified by scholars in the mid-1970s, the concept of food security rested on five assumptions:

1. Developing countries are able to maintain only inadequate reserves.
2. There is only a limited amount of available international emergency stocks of food.
3. Developing countries typically possess low levels of international financial reserves.
4. Developing states will assign high priority to preserving existing levels of imports deemed essential to their economic development.
5. Given the preceding assumptions, "low-income economies will be forced to respond to a short-term threat to food supplies by allowing damaging downward adjustments in consumption levels" (Green and Kirkpatrick, 1981).

Since internal food distribution networks tend to be inefficient and operate poorly in rural areas, any decrease in consumption levels at the national level will be aggravated for the rural poor, thus compounding the severity of the problem. Recognizing these facts, international scholars and officials began searching for short-term solutions to the problem which could be achieved through international cooperation.

Four alternative methods of stabilizing food supplies were proposed which relied on some level of national and/or international cooperation. The first proposal relied upon the generosity of net food producers. This scheme suggested that the food-surplus nations donate a portion of their surplus to countries experiencing drastic reductions in their food stocks. This alternative ignored two fundamental economic and political facts, however. Most net foods producers market their surpluses, which comprise a large portion of their export earnings. Donations of food stocks would thus reduce the amount available for export, therefore reducing income, and could possibly lower the price of such stocks on the world market. (This argument strengthened the position of proponents of a fifth solution, free trade). Second, many countries employ food as a foreign policy tool and would likely be reluctant to relinquish leverage over this tool (Adams, 1983; Murdoch, 1980).

Another two alternatives focused on a combination of national and international efforts. Many developing states and international agencies utilized improved technologies in agriculture to increase food production. Theoretically, this approach would result in long-term improvements in food security, but it was not an adequate solution given the short-term nature of the problem.

A third proposed solution, buffer stocks at the international and/or national levels, appeared to be the preferred solution in the 1974-1979 period. This method would allow states to stabilize their food stocks by establishing large supplies of reserves for use during shortfall periods. On the international level, these buffer stocks would be created and maintained by food-surplus nations for distribution during crises. In accordance with this manifest international mandate, talks began at the International Wheat Conference to establish an international buffer stock for grains, (Adams, 1983).

Despite the apparent preference for these types of arrangements among those experts advocating non-free trade solutions, it soon became obvious that the economic and administrative costs of such schemes were prohibitive, both nationally and internationally. Low-income developing states did not possess the capital to invest in large national buffer stocks of food, and the developed nations could not agree on the size, structure, distribution of cost, and prices of any international facility. Consequently, the talks collapsed in early 1979 (Adams, 1983).

While the IWC talks were proceeding, a number of scholars debated a new proposal for international cooperation to resolve the food security dilemma. This alternative concentrated on the international financial aspects of the problem: the inadequacy of international financial reserves in developing countries to meet the cost of purchasing food in shortfall years. Among the various plans proposed was a facility which would compensate experiencing a temporary decline in their cereal stocks or an increase in the price of their cereal imports. These funds would then be used by the recipient to purchase grain on the open market to supplement domestic stocks, thus stabilizing both the domestic price and volume of food stocks (Green and Kirkpatrick, 1981, 1982; Huddleston et al., 1984; Adams, 1983).

After the demise of the IWC talks, the Food and Agriculture Organization of the United Nations and the World Food Council turned their attention to the new proposal, now called the "Food Financing Facility." At this point, the

International Monetary Fund was suggested to administer the facility due to its experience in administering similar financing arrangements. With the cooperation of members of the IMF's staff, suggestions for the organization of the FFF were modified to accommodate IMF structure and procedures, and the WFC and FAO brought the proposal to the IMF in early 1980 (Adams, 1983; Huddleston, 1984; Green and Kirkpatrick, 1982).

Once the FFF entered the Fund, the FFF became entirely a Fund issue. "From this point on the IMF, with only a minimum of reliance on outside organizations, pursued its own analysis of the various problems" (Adams, 1983). As with the new facilities of the 1970s (e.g. the Trust Fund), the debate on the FFF was subsumes in the debate on larger issues such as the KIEU and the appropriate role of the Fund with respect to development issues. There was significant opposition within the Fund's membership to the expansion of the IMF's lending to include development financing. The U.S. had long espoused this position (Assetto, 1986; Wall Street Journal 14 June 1974, 7 January 1976; New York Times 27 September 1983) and consequently, became the leader of the opposition to the FFF.

Although the facility was to be available to all IMF member, the FFF was explicitly designed and intended to address the problem of food security in low-income countries (Green and Kirkpatrick, 1981; Huddleston et al., 1984; IMF, Survey, 9(19), 13 October 1980). The FFF, then, could be classified as a development facility. Consistent with its position on earlier development-oriented facilities, the U.S. opposed the inclusion of the FFF within the IMF; Germany, Canada, and Australia adopted the U.S. stance , while France and Japan supported the adoption of some form of the new facility (Adams, 1983; IMF, Survey, 9(19): 321, 13 October 1980). It is interesting to note that of the major opponents of the FFF, three were largely grain exporters.

Adams notes that contrary to expectations, the developing countries' support for the FFF was lukewarm (Adams, 1983). Indeed only two governors from Third World Countries (Sri Lanka and Bangladesh) spoke in favor of the facility at the IMF's 1980 Annual Meeting at which the WFC and FAO made their official request (IMF, Survey 9(19): 321; Adams, 1983; Green and Kirkpatrick, 1982). Adams attributes this to the fact that developing countries were much more concerned that the Governors approve the proposed 50 percent quota increase than adopt and FFF which would make available only a limited amount of additional funds

(1983: 557). The Group of 24, and advisory group to the IMF composed of delegates from the Fund's Third World members, and the Fund's Development Committee did advocate the adoption of the FFF, however (IMF, <u>Survey</u>, 9(19), 13 October 1980).

Ultimately, the United States and other developed countries approved the new facility, now termed the Cereal Imports Facility (CIF), though not without reservations. U.S. Under-secretary for Monetary Affairs (Treasury Department), Beryl M. Sprinkel, emphasized the limited nature of U.S. support of the facility, warning that the CIF was not "a step down the road" toward development financing in the Fund (<u>Wall Street Journal</u>, 22 May 1981). This acceptance of the CIT is attributable, in part, to the "pivotal" role of IMF Managing Director, Jacques de Larosiere (Adams, 1983: 558), but the attractiveness of the final structure of the facility must also be noted.

There were two forms of the CIF which were offered to the IMF for its consideration. One form called the institution of a separate food financing facility within the Fund which would later become linked to the existing Compensatory Financing Facility (CFF) through proposal of a joint quota limit. The second alternative required the integration of a cereals facility into the CFF with a slightly enlarged quota limitation. The second option was widely regarded as the least costly to the IMF and the international community and was the proposal which the IMF ultimately accepted. It was because the CIF could be "sold" as a relatively low-cost means of addressing the problem (Green and Kirkpatrick, 1981; Huddleston et al., 1984, Adams, 1983) and also satisfied developing countries' demands that the U.S. and its allies in the debate eventually approved the facility. As it was finally structured, the CIF did not require additional administrative and financial support from the IMF or its members, and it could be administered in a fairly mechanistic manner.

CIF Structure

The CIF which was instituted in May 1981, was designed explicitly to alleviate food insecurity arising when export shortfalls and/or sharp increases in the price of cereal imports threatened food stocks in a member country, particularly a low-income member (IMF 1981; IMF, <u>Survey</u>, 1981; Huddleston et al., 1984; Green and Kirkpatrick, 1981). Similar to the CFF, the CIF was designed to meet temporary shortfalls which were due to circumstances beyond the member's control (all details of the CIF are taken from IMF, Decision No. 6860-(81/81), 1981). Since Fund conditionality is predicted upon a member's internal policies and use of CIF facilities was limited to external conditions unaffected by those policies, conditionality associated with the CFF and CIF was deliberately low.

In practice, this was true until 1984. The original form of the CIF required only that the IMF be satisfied that "the member will cooperate with the Fund and in an effort to find, where required, appropriate solutions for its balance of payments difficulties" (IMF, 1981: 169) when its drawings on the facility exceeded 50 percent of its quota. A member was permitted to draw up to a maximum of 100 percent of its quota for either export shortfalls (declining export receipts) or import excesses (increased cost of imports); the total available to a member could not exceed 125 percent of quota. Amounts borrowed under the CFF, including the CIF, were considered in addition to the Fund's tranche facilities and thus, added to the absolute amount a member could borrow from the Fund. The CIF, however, only added 25 percent of the quota to the funds available to a member under the CFF.

In a 1984 decision revising the CFF, the IMF Executive Directors lowered the quota limits associated with the CFF to 83 percent of the quota and 105 percent of quota for combined drawings in the CFF and CIF (Decision No. 7528-(83/140; IMF, 1984: 137-138). In relative terms, this reduced the amount of assistance available under the CIF, although in absolute terms it did not negate the real increase in funds which resulted from the 50 percent increase in total Fund quotas which was enacted the next year. Nevertheless, the CIF was intended as a low-conditionality facility which added only a limited quantity to the amount that the member could purchase from the Fund.

Shortfalls and import excesses are calculated on the basis of trends established over a five year period. The CIF compensates the member to the extent that its exports fall below, or its import cost exceed, the level established by that trend line. Import excesses are calculated such that they are net of increases in export earnings. In simplified terms, the CIF appears as follows:

Shortfall=(trend export receipts - actual export receipts) + (actual cereal import costs - trend cereal import costs)   (Huddleston et al., 1984).

Initially the trend line was established for import cost excesses by computing the arithmetic average centered on the excess year, but this was later modified to the geometric average (Y1 x Y2 x Y3 x Y4 x Y5, see Green and Kirkpatrick, 1981); export receipts shortfalls are calculated using the geometric average centered on the shortfall year. This procedure requires that the member and the Fund estimate the figures for the shortfall/excess year and forecast the figures for the succeeding two years. If the calculations regarding shortfalls and excesses prove to be inaccurate, the member is required to repurchase (reimburse) from the Fund an amount equal to the error. In any case, a member must repurchase its drawing within 3 to 5 years.

CIF and Food Security

There are several benefits of the CIF as it is implemented in the IMF. Supporters of the facility claim that the CIF eases the foreign exchange constraints associated with food insecurity and allows low-income countries to bid competitively on the world market to replenish their stocks. The CIF also provides a cushion which permits members to stabilize domestic prices and an incentive to develop "rural procurement and marketing programs" (Huddleston et al., 1984: 61). In addition, the cereals facility, which applies only to shortfalls and excesses in cereals, could also mitigate the effects of decreases in non-cereal imports in shortfall years (Huddleston et al., 1984).

There are also several imperfections in the structure of the CIF which reduce its utility for food security. As constructed, the facility emphasizes the monetary value of shortfalls and excesses and does not address the problem of shortfalls in the volume of food. It is conceivable that due to rising prices CIF compensation would not alleviate food shortages in a country in terms of a sufficient amount of food (Green and Kirkpatrick, 1982). The restriction to cereals also limits the range of effectiveness of the CIF. In many countries, non-cereals account for over 40 percent of the diet of the populace; this is particularly true in Africa (Huddleston et al., 1984; Green and Kirkpatrick, 1982). In a sample of 40 developing nations in 1976 through 1978, Huddleston et al., (1984: 62-63) found that in 18 of these states, over 40 percent of the average diet consisted of non-cereals. For these states, therefore, the CIF is of limited utility.

It is also conceivable that the CIF would not serve the very constituency it was intended to reach, the lower-income, developing nations. First, linking to the quota the amount of CIF assistance available to a member restricts the funds at the member's disposal during a crisis. The absolute amount of funds cam make a critical difference to consumption levels of a poor country. Also since most of the lower-income developing countries have among the lowest quotas in the IMF, access to the CIF is biased in favor of countries with larger quotas and which are more likely to be able to afford additional food costs without assistance. The CIF, then, is NOT a development facility and could possibly be of very little help in maintaining consumption levels in low-income countries, if it is not beyond their reach altogether (Green and Kirkpatrick, 1982).

Limitations on the utility of the CIF for low-income developing countries are increased with the IMF's 1983 revision of the conditionality requirements for the CFF. Decision No. 7528-(83/140) of the IMF's Executive Board describes the new conditions applicable to drawing on the CFF and CIF in more specificity than the original 1963 and 1981 decisions, respectively (IMF, 1984: 137). The term "willingness" is now defined in specific terms as:

> "a willingness to receive Fund missions and to discuss, in good faith, the appropriateness of the member's policies and whether changes in the member's policies are necessary to deal with its balance of payments difficulties. Where the Fund considers that the existing policies of the member in dealing with its balance of payments difficulties are seriously deficient or where the country's record of cooperation in the recent past has been unsatisfactory, the Fund will expect the member of take action that gives, <u>prior to submission of the request for the purchase</u>, a reasonable assurance that policies corrective of the member's balance of payments problem will be adopted" (IMF, 1984: 137; author's emphasis).

Above the 50 percent of quota level, stricter conditionality similar to that applied purchases made in the upper credit tranche in the Fund, would be applied. As defined by the Fund, cooperation is considered to be "the existence of and broadly satisfactory performance under an arrangement with the Fund, or the adoption of such an arrangement at the time" of the request (IMF, 1984: 137).

Thus, a facility which was designed to ensure easy and swift access to assistance is not subject to conditions which permit the Fund to scrutinize a member's domestic policies <u>before</u> a purchase is even requested. According to the requirements of the CIF, however, a shortfall must a be beyond a member's control in order for that member to qualify for CIF assistance (IMF, 1981: 1). In practical terms, this measure means that in seeking a purchase under the CIF; the member has become responsible for conditions, <u>which by definition</u>, are beyond its control.

In strictly technical terms there is also some doubt about the reliability of the measurement and forecasting techniques the Fund employs to determine the extent of the shortfall, and the degree to which that shortfall is die to external or internal conditions.[3] Inaccuracies in these figures could cost the borrowing member vital funds in terms of lower shortfall estimates or the imposition of stricter conditionality.

In addition, CIF purchases, which under the previous rules required only a month between request and extension of assistance (Dell, 1985; Huddleston et al., 1984), may now be delayed by the lengthy process necessary to complete the Fund's review of the member's policies. This may pose an undue burden to countries already on the brink of serious food shortages. As Huddleston et al., stated:

> "Years of food shortages are not the appropriate time to delay assistance while domestic policies are investigated. The test for borrowing under the facility should not go beyond the determination of need in the sense of an actual or expected increase in the food import bill from causes beyond the government's control" (1984: 68).

Further, the imposition of increased conditionality on CIF loans may actually inhibit states from seeking such assistance because of the added expense and the potential political costs of an IMF review. These costs are relatively greater for the lower-income countries which would seek CIF assistance.

Usage of the CIF by IMF Members

As shown in Table 10.1, IMF members have made little use of the CIF in the 5 years since its exception. Although all Fund members except Morocco (see below) were eligible for

Table 10.1
Purchases in the CIF, 1981-1985
(in millions of SDRs)

Year	Member	Amount of CIF Purchase	% Quota	Amount of CFF Purchases by Other Members	Number of CFF Purchases
1982	Korea	106.2	41.5	1,280.2	21
	Malawi	12.0	42.1		
	Morocco	236.4	105.0		
		236.4			
1983	Bangladesh	71.2	31.2	3,596.3	26
	Kenya	60.4	58.3		
	Malawi	12.2	42.8		
		143.8			
1984		0.0		1,180.0	13
1985	Bangladesh	55.0	19.0	784.0	9
	Ghana	58.2	28.0		
	Jordan	57.4	77.6		
	Korea	279.7	60.4		
	Malawi	13.8	37.0		
		464.1			

Total CIF = 962.5
Total CFF = 6,840.4

Source: IMF, Annual Report, 1981-1985, (Washington, D.C.: IMF).

some degree of CIF and/or CFF assistance during this entire period, only seven members have made purchases under the CIF component of the CFF, for a total of SDR 962.5 million (in 12 purchases). Of that total, the Republic of Korea alone accounts for SDR 385.9 million. For the same period, members made 69 purchases totaling SDR 6840.41 million under the regular Compensatory Financing Facility (IMF, 1981-1985). During this period 1981-1985, no member borrowed simultaneously from the CIF and CFF, and only one member, Morocco, borrowed an amount equal to the limit placed on it by its quota.

Total purchases under both aspects of the CFF from 1982-1985 were SDR 7802.91 million. CFF purchases, therefore, have constituted only 12.33 percent of total CFF drawings. And, of the seven members which made purchases under the CIF, only three qualify as "Most Seriously Affected" (MSA) members. Clearly, the CIF has not been of major significance in the attempt to solve the food security problem.

These results seem to confirm the argument detailed in the previous section which states that given the nature of innovation in the IMF, the CIF would likely be of limited utility to low-income, non-cereals dependent states. Members have been hesitant to sue the CIF, and those which have chosen to do so are not necessarily MSA's. In fact, the largest user of the CIF has been South Korea (40 percent of CIF purchases since 1981), widely considered to be a prosperous developing country (or NIC).

The absence of any CIF purchases from April 1983 through April 1984, a period when several Africans states began experiencing severe food shortages, also supports this conclusion. African diets are typically not cereal-dependent and of the 18 African states seriously affected by the 1983-1984 food crisis (<u>Newsweek</u>, 24 November 1984: 54), only Morocco and Kenya took advantage of the CIF, and neither of these borrowed more than 60 percent of their quota limits. In addition, only four of these countries took advantage of the CFF during the same period, although all 18 countries were eligible for both CIF and CFF assistance: Niger, SDR 24 million; Sudan, SDR 39.1 million; Zambia, SDR 131.2 million; and Zimbabwe, SDR 56.1 million. While access to less costly sources of food assistance can explain this apparent reluctance to use of the CIF, it is certainly possible that the lack of purchases in the CIF in 1984 may also be attributed to the increased conditionality requirements of the 1984 revision of the CFF.

CONCLUSION

The Cereals Import Facility was established by the International Monetary Fund at the urging of the WFC and the FAO to ameliorate the international food insecurity problem, especially in low-income developing countries. The CIF was formulated as a part of the Fund's existing Compensatory Financing Facility which compensates IMF members which are experiencing temporary shortfalls in their export receipts. Thus, the structure of the CIF paralleled that of the CFF, considering cereal imports as negative exports in the calculation of shortfalls.

The CIF as it is presently constituted suffers from several defects which serve to inhibit its use by low-income developing countries and therefore, reduce its efficacy. The linkage to quotas of the amount of CIF funds for which a member is eligible reduces the absolute amount of assistance upon which that member can draw in periods of emergency. The increased conditionality of CIF drawings also inhibits borrowing from the facility due to higher political and economic costs. Due to the cereals focus of the facility, some low-income countries which do not rely heavily on cereals in the composition of their diet also receive reduced benefits under the scheme.

The conditionality which is not applied to the CIF, as well as the CFF, makes the facility most unattractive to developing states. Compared to both multilateral and bilateral food aid, the economic and political "strings" associated with IMF conditionality make the CIF more costly, particularly in political terms. Food aid tends to be donated in the forms of grants, not loans which must be repaid, and such aid usually does not require policy alterations in return. Conversely, CIF food assistance requires repayment, albeit at less than market interest rates, and may subject the borrower to undesirable, and potentially politically unsettling, policy conditions. These costs may appear even higher when the fact that the funds obtainable through the CIF are not usually sufficient to meet the entire need, is taken into consideration. Indeed, more generous and less costly sources of food assistance easily explains the infrequent usage of the CIF, and therefore, it is not surprising that developing countries have gravitated toward source of food assistance outside the CIF.

The structure of the CIF and its practical consequences are predictable given the nature of innovation in the IMF described in the IMF, as described at the beginning of this chapter. Unlike the oil crisis, the food security crises of 1972-1974 and 1983-1984 were not systemic crises and therefore, did not seriously affect the Fund's most powerful members. The legitimacy and survival of the IMF itself was not at stake; indeed the Fund was almost immediately regarded as the organization most capable of administering an international food financing facility. There were no incentives for the IMF to commit its reputation and resources to the development of a solution to the food security problem. The fact that the Fund did adopt the CIF was due primarily to the low cost of the facility and the political capital gained by the Fund and the developed countries by addressing the problem in some fashion.

The CIF was not a significant departure from standard Fund procedures. The Fund's conceptualization of the problem permitted it to cast the new program within the structure of the existing CFF and to administer it in much the same manner. Thus, the CIF was not only low-cost (to the Fund) in terms of actual financial assistance but also in terms of administrative expense. This is consistent with the earlier argument that the Fund tends not to innovate unless compelled to do so by a crisis which involves its most influential members. This suggests that in order to succeed, future innovations introduced in the Fund during non-crisis periods must incur no significant additional expense to the organization and be couched in terms familiar to the Fund.

The CIF, should be considered in the context of a symbolic response to the food security needs of the developing countries. Because of its very nature, attempts to revise the CIF in directions which would make it more appealing to developing countries are likely to fail. Unless political conditions within the IMF and in the broader international distribution of power change significantly, the CIF will remain of limited value to the very states which need it most.

NOTES

1. For more complete treatment of this argument, see V.J. Assetto. 1986.

2. See Assetto, 1986: 2-9.

3. Dell, 1985, and personal communication with Dr. Harvey Cutler, Assistant Professor of Economics, Colorado State University, Ft. Collins, CO.

REFERENCES

Adams, Jr., Richard (1985). "The Role of Research in Policy Development: The Creation of the IMF Cereal Import Facility." World Development 13 (7): 549-563.

Assetto, Valerie J. (1986). "Crisis and Innovation in the IMF." Paper presented at the 1985 Annual Meeting of the International Studies Association, March 25-29, 1986, Anaheim, CA.

Dell, Sidney (1985). "The Fifth Credit Tranche." World Development 13 (2): 245-249.

Green, C. and C. Kirkpatrick (1981). "Insecurity, Food Financing and the IMF." Food Policy 6 (3): 135-146.

_____ (1982). "The IMF's Food Financing Facility." Journal of World Trade Law 16(May/June): 265-273.

Huddleston, Barbara, D. Gale Johnson, Shlomo Reutlinger, and Alberto Valdez (1984). International Finance for Food Security. Baltimore: The Johns Hopkins University Press.

International Monetary Fund (1981-1985). Annual Report. Washington, D.C.: IMF.

_____ (1981-1985). Survey. Washington, D.C.: IMF.

Murdoch, William M. (1980). The Poverty of Nations: The Political Economy of Hunger and Population. Baltimore: The Johns Hopkins University Press.

Newsweek (26 November 1984). "A Continent in Mortal Danger."

New York Times (27 September 1983), IV 6: 1.

Wall Street Journal (1981-1985). Various Issues.

11

Who Governs the Rome Food Agencies?

Ross B. Talbot and A. Wayne Moyer

The purposed of this working paper is to analyze the organization of power in the four world food organizations headquartered in Rome, Italy. We refer specifically to the Food and Agriculture Organization (FAO), World Food Council (WFC), the World Food Program (WFP) and the International Fund for Agricultural Development (IFAD).

Our basic premises are: (1) this is a world of nation-states, each claiming to be sovereign; (2) a nation-state joins and participates in an international organization (IO) in pursuit of its own perceived national interests; (3) in order to achieve this objective, the member-state will usually form a coalition with other states which have closely identical interests; (4) policy outcomes within the IO will be determined through the negotiations carried on within and between these political coalitions. Political scientists have engaged in countless arguments over the definition of power but will arbitrarily use the one advanced by Thomas Hobbes: Power is the present means to achieve some future apparent good. The means will primarily be wealth (money, resources and technology) and votes; in an IO each member-state has one vote, although Orwellian logic generally prevails (i.e., some nation states are more, or less, equal than others.

The humanitarian factor must also enter into these calculations relating to the use of power. There is a deep-routed stream of idealism in Western culture; we will eagerly and seriously engage ourselves in a war on hunger, under certain circumstances. Starvation and famine are cultural taboos within rich nations; however, malnutrition is less of a social-psychological inhibition to rich nations in their relations with developing nations. Nevertheless, there is an obverse side to this value proposition. Food is obviously a basic human need, but too much food enters into the political calculations of food-surplus nations with at least as much potency as does too little. This complicates policy making for wealthy, and (usually) food-surplus, nations. Their capabilities for supplying food to needy nations are evident; but the motives for doing so are complex. Food aid comes easy when farm surpluses are a burden to the national economy. However, when the wealthy nation enters a period of rising food prices then the concern for meeting human needs in poor nations tends to be seriously moderated. Then political realism takes a commanding role; domestic economic interests (i.e., stable food prices) must be served, first.

Although we will have to examine this matter superficially, it must at least be noted that power always has a historical dimension. The structure of world power is continuously in a state of flux and change, and these international organizations are both a historic and contemporary resultant of this struggle for power among nations. The FAO became an international organization in the 1943-1945 period. It was a creature of the Allied Powers, meaning predominantly the U.S., United Kingdom, and Canada, with the exclusion, by choice, of the Soviet Union. Third World nations, excluding Latin America, were nearly all colonies at that time. FAO members originally numbered 34; today there are 158, and the structure of power has been significantly altered.

The World Food Program is a product of the 1960-1963 period; sponsored by the U.S. Government, abetted by the FAO, supported by Canada and Australia, and somewhat reluctantly agreed to by the new and suspicious developing nations. Today, the European Economic Community has become a major contributor and Third World nations have acquired a power status nearing equality within the WFP. Both the World Food Council and the International Fund for Agricultural Development were the creations of the World Food Conference, which was held in Rome in November 1974, at a time when a worldwide food shortage seemed to be on the immediate horizon. But their configurations of power differ decidedly from each other, and from FAO and WFP. The World Food Council is primarily a policy innovator, a minor broker in super-power politics, with the U.S., Soviet Union, and China as continuing members. IFAD has a unique configuration: OECD (Organization of Economic Cooperation and Development)---OPEC (Organization of Petroleum Exporting Countries)---and Group of 77 (Third World Nations). Process follows structure, by and large and, as we will indicate, this means that the issue of "who governs?" must often be answered differently over time. Although we will be unable to develop this proposition in depth for the 1945-1985 period, we will endeavor not to understate in our analysis the historical dimension which is involved. The past is a prelude to, but different from, the present.

WHO GOVERNS? - AN INTERNATIONAL TRIANGLE (QUADRANGLE) MODEL

We will speak to that question, in turn, for each international organization. However, we were unable to conceptualize a single model which would explain both structure and process for all four organizations. Conceptually, an international triangle of power works fairly well for FAO and WFP. One side of the power triangle would be the OECD member-states (i.e., the nations of Western Europe and North America, plus Japan, Australia, and New Zealand) who are, by far, the principle funding actors; a second side would be the Third World nations, somewhat loosely organized into a political coalition called the Group of 77 who are the principal recipients of this funding; with a third side being the managerial staffs (the bureaucracy) of FAO and WFP, which have the crucial roles of policy innovator and mediator relative to promoting mutually acceptable agreements between those who fund and those who receive (Keohane, 1984: esp. Ch. 4-6). But the triangle-of-power concept has to be restructured into a quadrangle in order to explain the workings of IFAD: OECD, OPEC, Group of 77, and the IFAD bureaucracy. And for the World Food Council, the Soviet Union and the Eastern Europe bloc has to be included, and OPEC becomes a part (unrealistically, in economic terms) of the Group of 77.

These schematic arrangements are useful in explaining how the world food organizations function. There is, to some extent, a commonality of interests. OECD nations (the North) have political, economic and cultural interests in assisting developing nations (the South) to emerge from conditions of poverty; the developing nations, obviously have the desire to do so. but claim that a prerequisite is abundant development assistance. On the other hand, this agreement on goals and objectives often does not translate into an agreements on means and methods. North and South need each other, but the latter's demands are insatiable, while the former believes that the resources available for redistribution are limited. The respective international bureaucracies function as a kind of influential prime-mover, inclining toward finding policies and means to meet the demands of the South, but understanding the limitations of their power to make a claim on the wealth of the North. Moreover, these bureaucracies have vested interests of their own. The result is compromise, incremental change which is presumed to be an optional arrangement by the major political actors at that particular moment.

Also, none of these power configurations take sufficiently into account three other important considerations. One is the influence on the world food agencies of the United Nations itself and some of the other IO's---specifically, the United Nations Development Program (UNDP), the World Bank and the regional banks, and, to a lesser extent, such organizations as the World Health Organization (WHO) and the International Labor Organization (ILO). Secondly, the NGO's (non-governmental organizations) such as CARE, Catholic Relief Service and Church World Service, are secondary actors, but not without influence, in the policy-making of each of the world food organizations. They are effectively organized, politically and functionally, in North America, Western Europe and Australia, and are closely associated with an international structure in Geneva--The International Council of Voluntary Agencies. Thirdly, in each international organization there is a director-general (or an executive director or president) and he is certainly first and foremost in the bureaucratic hierarchy; jealous of his authority, suspicious of competitors, authoritarian (in varying degrees) in his leadership style, although dependent on the imaginative insights, analytical expertise, personal and corporate loyalty, and administrative skills of the IO's professional and technical staff. Moreover, conflicts of power, always prevalent and occasionally of some magnitude, exist within and between the bureaucracies of these world food agencies. At least, such was and is their past and present condition, and we would anticipate little, if any, changes in the foreseeable future.

In any respect, we could not develop a single model that would have explanatory power for all of these agencies, and the two variations we have offered are not without their defects, which we will later elaborate.

WHO GOVERNS? A DESCRIPTIVE EXPLANATION

In this section we will attempt an answer for each of the world food agencies to our central question: Who governs? Our presentation cannot be comprehensive but we will provide an overview based on the following outline. First, there will be a brief look at the constitutional-political basis of power of each IO. Those who study IO's tend to overlook the constitutional basis from which the agency must proceed; if this legal framework is abused or violated it is quite likely that serious political conflict will ensue, and little will be accomplished until those constitutional issues are resolved. Second, we will explain the budgetary-financial basis of power of each organization. An IO does not have the power to tax; even less does it have the power of the sword (i.e., exacting penalties on the negligent or recalcitrant member-state). But the question of: who pays?-who receives? always generates fundamental political and economic issues, the answers to which tell us much about who gets what, when, and how. Third, we will sketch out the policy-making process, and so by briefly explaining, where applicable, the internal power relationships which occur at the various stages of the policy cycle. We will use the conventional, five-stage policy cycle: agenda building, formulation, legitimation, implementation, and evaluation.

What eventuates, in terms of policy decisions and their implementations, is what a reasonably knowledgeable person would suspect, based on the power models we discussed earlier. That is, the policy process in the international organizations is one of compromise and mutual accommodation, resulting usually in incremental change. Conflicts of interest and ideology do occur, of course, and the competing demands are decided on in an environment in which the principal actors must search for a consensus. One of the definitions of power offered by Keohane and Nye (1977: 11)---"...The ability of actor to get others to do something they otherwise would not do (at an acceptable cost to the actor)."---has come to have a special meaning in these IO's. That is, the Third World nations have the votes, but not the resources; conversely, the have---nations have their own interests which not be served by withdrawal or persistent "stonewalling". The power of the actors (member-states and bureaucracies) is so limited that, confronted, with interests both common and conflicting, there is really no viable alternative other than a search for a mutually-acceptable consensus.

WHO GOVERNS THE FOOD AND AGRICULTURE ORGANIZATION?

The Food and Agriculture Organization of the United Nations was created in 1945 as one of the autonomous specialized agencies of the United Nations system whose purpose was to promote cooperation among nations and encourage action on common problems thus contributing to world peace. FAO actually predated the formal establishment of the United Nations, and was given a broad mandate under its constitution. It has four purposes:

1. Raising levels of nutrition and standards of living under the jurisdiction of the member governments;
2. Securing improvements in the efficiency of production and distribution of food and agricultural products;
3. Bettering the conditions of rural populations;
4. Contribution to an expanding world economy and striving to assure freedom from hunger (Phillips, 1981: 9).

According to the U.S. Senate Select Committee on Nutrition and Human Needs (1976: 20) the organization which emerged reflected a compromise between those who wanted a strong action-oriented agency to foster agricultural development and those who wanted a more limited fact gathering and advisory agency.

FAO was established with an organizational structure fairly common for intergovernmental organizations. The supreme governing body is the Conference which holds regular sessions biennially and elects the Director-General. Each member-government may send one delegate to the Conference along with alternate and staff, and each member has one vote. The Conference acts on application for membership, decides on the budget level and approves the FAO programs of work which is drawn up by the Director-General and FAO Secretariat. The Conference also decides the scale of member contributions to FAO. Assessments have generally been made in accordance with the UN formula with members contributing a percentage of the FAO budget proportionate to the relative size of Gross National Product (GNP).

Another function of the Conference is to elect member countries to the Council which meets between Conferences and serves as a second-level governing body. The membership of the Council has grown to 49 members. The U.S. has always been a member. Much of the substantive work discussed by the Conference and Council is carried out by committees. There are seven standing committees dealing with program, future, constitutional and legal issues, commodities, agriculture, forestry and fisheries. The Conference and Council also carry on their work through a number of ad hoc bodies such as the Committee on World Food Security and the Commission on Fertilizers.

FAO has grown very significantly since its inception. Its budget in the 1946-47 biennium was $8.4 million. For the 1986-87 biennium, total finding is $1.1 billion. In FAO's early years, prior to decolonization, FAO was dominated by the Western industrial powers who took a rather restricted view of what the organization could accomplish. Primary emphasis was given to making technical studies, collecting and publishing statistics, conduction conferences, establishing technical commissions and dispatching occasional field study missions. In this period, the FAO budget grew slowly.

Independence for the former colonies led to a significant increase in the membership of FAO in the 1950s and 1960s and a shift in the voting balance to give the developing nations a significant majority in the Conference. New pressures were generated on FAO to change its focus from information gathering and dissemination to field activities in support of Third World agricultural development. An active field program was developed by FAO, funded by extra-budgetary resources contributed primarily by UN agencies and national governments. However, the transition proceeded slowly. Change was resisted by the FAO bureaucracy and resistance developed, and continues, among the industrial nations to FAO becoming primarily a development agency.

The early 1970s saw the development of a world food crises caused by global drought occurring in the context of Third World population explosion. FAO was severely criticized for not anticipating the crisis and for not having done enough to stimulate Third World agricultural development and thus limit the effects of the famine. The general dissatisfaction with FAO was an inportant factor leading to the 1974 Rome World Food Conference, which was held under UN rather that the FAO auspices. An important aspect of this meeting at the UN was that the Soviet Union was (and is) not a member of the FAO. New international food organizations, independent of FAO, were established to accomplish missions which it was thought were not well carried out by FAO.

The Rome World Food Conference galvanized change in FAO. The sense of crisis forged a new consensus of OECD and Group of 77 countries, which led to the election of a dynamic Director-General, Edouard Saouma, committed to Third World agricultural development and widespread support for such reforms as the development of an executive system of country representatives, the establishment of a Technical Cooperation Program with quick grants for short term development needs, increased support for agricultural investment, and a considered effort to shift personnel resources from headquarters to the field. The crisis atmosphere facilitated mutual adjustment along with a rapid growth of FAO's budget and extra-budgetary resources.

As the world food crisis receded in the late 1970s, the consensus between OECD and the Group of 77 weakened and FAO's budget growth and innovativeness slowed down perceptibly. The U.S., concerned about an increasing balance of payments deficit, among other reasons, began to exert cost containment pressures. Since 1983, FAO has operated with almost a no-growth budget.

The FAO policy process is a cumbersome one with a strong tendency toward incrementalism as one would expect from an organization with 158 member governments and 6,600 staff. The program of work is put together in the various divisions of FAO, and the Director-General makes the final determination of priorities after consultation with the staff and member countries. Director General Saouma has a decisive style and dominates the policy process. He does have to walk a very fine line between meeting the demands of the Group of 77 countries for expanding FAO programs without offending the OECD countries, which still provide the bulk of the funding. The Conference, in practice, is unwieldy and does not play a major role in planning or determining priorities. Its major function appears to be a ratification of the decisions reached by the Director-General and staff. The Council is more involved than the Conference in FAO substantive matters, primarily as a sounding board for new proposals, but still tends to defer to the Director-General and the Secretariat.

FAO is severely constrained in its policy autonomy in that its resources are primarily technical not financial. Unlike the World Bank and IFAD, FAO cannot carry out development projects, but can only provide technical support. Hence, FAO activities must remain very closely tied to the projects funded by the major lending agencies. FAO's dependence on extra-budgetary funding provides another constraint, with only about 38 percent of total spending provided by the regular assessed budget (Kriesberg, 1984: 48). The remainder comes from a variety of grants and trust funds provided by international organizations and countries, and FAO pretty much must do what the donors want to retain access to this funding.

The Director-General gains the necessary support to legitimate his policies in a number of ways. He can use his discretionary authority to distribute Technical Cooperation Program funds to gain support from Third World governments. He also has significant power to make staff appointments and can exert significant leverage on member governments by adroit use of his patronage power, at some apparent cost to the general technical competence of the FAO staff. He maintains influence internally through control of promotions and by placing his own people in important positions.

FAO can implement its policies only to the extent that it has the support of the nations where it operates. Our sense is that FAO has gained a freer band to function in Third World countries as the perceived need for agricultural development has increased. A more serious problem is the sensitivity of many governments to criticism. Hence, FAO has been slow to develop meaningful evaluations and in communicating the results of such evaluations as it made. since the World Food Conference, FAO has developed a more rigorous evaluation process, though these assessments are still not generally available to outsiders.

Another obstacle to effective implementation is that FAO programs need to be coordinated with the activities of other international organizations and of national governments. this was very difficult prior to the Rome World Food Conference when most of FAO's personnel were headquarters based. Director-General Sauoma has alleviated this problem somewhat with his system of country representatives, which now extends to more than 74 countries. These country representatives, under the direct control of the Director-General, are responsible for seeing that FAO intentions are carried out, an in coordinating FAO activities with those of host governments and other international organizations.

In a general way, FAO has shown itself effective in adjusting to shifting international priorities. It responded effectively to the call of the Rome World Food Conference for an increased emphasis on agricultural production and investment and for the development of an agricultural early-warning system. It has also moved to extend its efforts to the small farmer by sponsoring the 1979 World Conference on Agrarian Reform and rural Development (WCARRD), and by emphasizing small farmers in its recent programs of work and budget. It has also given emphasis in recent years to food production in Africa, the region with the most serious problems. One should note, however, that it is not possible to measure the increased commitment to Africa and to agricultural reform by analyzing FAO budget figures (Moyer, 1986). One can surmise that the political balance is so delicate that it is difficult to make any significant policy changes in the absence of real budgetary growth.

WHO GOVERNS THE WORLD FOOD COUNCIL?

The World Food Council is a product of the World Food Conference 1974. Resolutions XIII, which was later ratified by the UN's Economic and Social Council and then the General Assembly, designated the WFC as the highest political institution in the UN system dealing with world food policies and problems. It is not an operating agency, "... but rather a forum and mechanism for initiating ideas and for reviewing the work of other international organizations with operating programs" (Kriesberg, 1984: 121).

The construction of the 36-member Council gives some political leverage to the Group of 77, whose members will occupy twenty-five of the Council seats. But the Council has no funds for projects or programs, only ideas and proposals. Its operating budget amount to less than two million dollars, sufficient to pay for an almost minuscule bureaucracy, consultants, and internal programming.

Actually the Council exists because, in the minds of many of the delegates at the World Food Conference, the Food and Agriculture Organization failed to fulfill its principal mission (i.e., to foresee and resumably then to prevent the world food crisis of 1972-1974). This is likely an unfair and misdirected charge, but the FAO has suffered from, and has become embittered by, this perceived failure. The relationship between the two organizations, particularly at the highest levels, continues to be one of studied, brooding, and mutual incompatibility.

Fortunately, in our opinion, the Council's Executive Director and (to a lesser extent) its elected President have provided the Council with dynamic and imaginative leadership. Its annual agenda has been dictated to a minor extent by desires made explicit at a previous Council session, but much more often Council proposals concerning "what should be done" to alleviate world food problems are the brainchild of the Executive Director, his staff, and their consultants. Searching for an exposing ways and means to improve food production, enhance world food security, and increase the flow of food aid constitute the Council's central objectives. The annual formulations of proposed strategies and programs are contained in a set of documents drafted by the WFC bureaucracy, which herefore have first been considered at the Prepatory Meeting, and then ratified, largely without revision, by the Council at its annual session, usually held in June each year, in a different continent. The Group of 77, through the device of the Prepatory Meeting, has endeavored to gain control of the Council's agenda and to considerably, if not radically, influence the proposals, but the Executive Director has managed to maintain control. Indeed, in 1986 the Preparatory Meeting was simply abolished.

The Council is not an implementing agency, and has no formal mechanism for evaluating the activities and programs of the other world food agencies. In establishing the Council the World Food Conference envisaged high-level, policy-discussing annual meetings, composed not only of ministers of agriculture but also with some representation of those from finance, planning, and development. But the 1974 world food crisis began in 1975-1976 to slip down from its status as high politics on the world scene to a kind of middle-level position. Gradually, there developed the realization that the world food problem, at its core and conscience, was actually one of poverty. A massive redistribution, from rich to poverty nations, would have to take place if this condition was to be attacked in a serious, concerted manner. The political environment simply did not (and does not) exist for that kind of revolutionary policy making to take place.

The World Food Council does not, and cannot govern; it endeavors to persuade (to use a prominent and current example) the Third World Nations to formulate and implement a national food strategy which will, over extended time, enable that food-deficit nation to become food self-reliant. And simultaneously the Executive Director and his staff have persisted in their efforts to persuade the OECD nations that they, in turn, must support---financially and through non-protectionist trade policies---those Third World nations who are endeavoring to implement a national food strategy.

Not only does the World Food Council seek to be creative and innovative, but it must also be concerned with searching for the means whereby ideas can be transformed into effective policies and programs. Over the last decade much has been heard of the concept of political will: i.e., the international community, could overcome the scourge of poverty---and the problem of food, shelter, ignorance and illness which follow in its wake---if the OECD and Group of

77 member-states displayed the political will to do so. Ideally, perhaps so, but the peroration has been essentially self-denying. This world, we would argue, is composed of may interdependent nation-states, but they do not perceive of themselves as constituting a world community. From those according to their ability to those according to their need is not the dominant theme in world politics---yesterday, today, or in the foreseeable future. The World Food Council endeavors to function as a policy innovator and political broker between those who have and those who have not. In that pursuit the Council has been successful to the extent that a North-South dialogue continues on major issues relating to agriculture and food, and many of these issues have been either initiated or enlightened by the endeavors of the Council.

WHO GOVERNS THE WORLD FOOD PROGRAM?

Viewed constitutionally, the World Food Program is a voluntary agency built upon a FAO Conference resolution passed in November, 1961, and subsequent UN General Assembly resolutions enacted in 1961, 1965, and 1975 (The World Food Program, 1978: 55). The economic reality was that the origin of the WFP came about because of the growing burden of over-bountiful farm surpluses, especially in the U.S. The political reality was two-fold: (1) the FAO bureaucracy had devised a program whereby surplus agricultural commodities of rich nations could be utilized in Third World countries for rural development projects, welfare programs, and food-disaster emergencies; (2) the Kennedy Administration---more explicitly, the Food For Peace director George McGovern---seized the opportunity presented by the FAO scheme and proposed the establishment of a World Food Program, financed predominantly by U.S. agricultural commodities and dollars.

Over time the WFP has become a unique type of voluntary agency. In essence, it is financed primarily (80-90 percent) by OECD nations who contribute food, cash and services for the support of WFP-sponsored development and disaster relief projects. Every two years a pledging conference is held in New York. In 1961, the pledges are valued at $100 million. By the end of 1985, just over one billion dollars had been pledged for the 1985/86 biennium, although this amount was some 25 percent short of the established target of $1.3 billion.

This fund of food, cash and services is spent on projects, both development and emergency, in Third World nations. Development projects are broadly of two types---primarily for agricultural and rural development (e.g., land rehabilitation, soil conservation, irrigation) and to a lesser extent for human resources development (nutritional support for mother, their infants, and primary school children). In 1985, these development projects were valued at $642 million. During the 1980s, emergency food commitments (disaster induced by humans, drought, natural catastrophes) have been on the increase, and in 1985 their cost came to over $225 million.

Because the finding is voluntary the policy-making process must be two-sided. Funding must first be volunteered, essentially by OECD countries; the expenditures are completely within Third World countries. Herein lies the international iron triangle, with the WFP bureaucracy occupying a prominent corner through its influence in the determinations of the project cycle. That is, projects are initiated and formulated by a Third World nation, with the assistance of the WFP Field Representative, alone with his/her staff. The legitimation stage take place in the CFA (the Committee on Food Aid Policy and Programs). In a sense, CFA performs a legislative function. Its 30 members-composed of OECD and Groups of 77 members, in approximately equal numbers-discuss each project, and occasionally will influence the WFP staff to modify the terms of a project proposal. Small projects, less that $1.5 million in value, may be approved by the Executive Director, then reported and justified to CFA. Adequate food resources are available so projects are rarely denied, at either the formulation or legitimation stage. (Funds, however, are usually in short supply.) The implementation of an approved project is primarily the responsibility of the recipient Third World nation, although the WFP is gradually becoming more involved in the transportation and distribution functions within that nation. The WFP has been a pioneer agency in the use of project evaluations, which must be reported periodically to CFA, and a final evaluation report is required after a project has been completed.

Consequently, to answer the question of who governs the WFP the response must be somewhat complex. On the input side, the OECD nations have to make the major financial commitment, although agricultural surpluses in North America, Western Europe, and Australia have meant that the opportunity costs of food aid have been quite low in recent years. On the output (project) side, the WFP staff and the proposing/recipient Third World nations play the dominant roles. Internally, too, catastrophic events---such as Kampuchea, the Sabel, and the Ethiopian food crises---are an enogeneous variable which, in reality, mandate a positive decision. The WFP staff desires to be viewed as primarily a rural development agency, but Third World disasters (man-made and natural) impose an increasing and recurring responsibility. In 1985, for example, WFP associates to Ethiopia cost over $211 million for 14 development projects and almost $70 million for 20 emergency operations.

As mentioned earlier, this triangular power configuration also has to be modified, at least marginally, because of the desires and influence of the NGO's and other UN specialized organizations. But, in general terms and discounting the exceptions to the rule, we believe that the triangular model has significant explanatory power.

THE INTERNATIONAL FUND FOR AGRICULTURAL DEVELOPMENT (IFAD)

The World Food Conference's Resolution XIII states that "an international Fund for Agricultural Development should be established immediately to finance agricultural development projects primarily for food production in the developing countries" (WFP, 1978). Implicit, too, in this resolve was that IFAD's projects should be aimed predominantly at the rural peasantry who live in conditions of abject poverty---the poorest of the poor," to use the vernacular phrase. The concurrence of the UN General Assembly soon followed, and the UN Secretary-General proceeded to initiate a search for voluntary pledges to finance IFAD's operations. Funding has constantly been one of its principal concerns. Probably no international organizations has been accorded as such a positive, indeed enthusiastic, public press as has IFAD, but none has been plagued by so many obstacles in the search for funds.

Resolution XIII also set forth, in somewhat vague terms, the constitutional structure of IFAD: A Governing Board representing "contributing developed countries, (i.e., OECD), and potential recipient countries (i.e., Third World nations" (WFP, 1978). Moreover, this representation was to be selected so as to ensure "regional balance" and an "equitable distribution". After some difficult negotiations, a set of Articles of Agreement was drafted, and opened for signature by willing-nation states. But a serious controversy between OECD and OPEC member-states over funding was resolved only after difficult negotiations, so

IFAD could not commence operation until November 30, 1977 (IFAD, 1977 and 1978). The Articles specified the establishment of a Governing Council (all signatory countries) as a tripartite Executive Board with each part having one-third of the 1,800 votes. And according to Section 6(b), "decisions of the Executive Board shall be taken by a majority of three-fifths of the votes cast (except for voting for suspensions or amendments), provided that such majority is more than half of the total number of votes of all members of the Executive Board."

The realities of power within IFAD are quite different, however, from the constitutional requirements. Presently, some 141 nations are voluntary members of IFAD, and constitute its Governing Council. The Council meets annually; its authority is decidedly limited, but the Council formally elects a President (the first was from Saudi Arabia, the second from Algeria), who in turn selects a Vice President (thus far, an American). However, political power actually lies with an Executive Board of 18 members: six in Category I (OECD, six in Category II (OPEC), six in Category III (Group of 77). Initially nearly all of IFAD's small bureaucracy were on secondment from other international or national aid agencies, although this is much less the situation today. The reputation of this bureaucracy has been consistently high: dynamic, experienced, resourceful, innovative.

IFAD's conundrum has been: how to secure finding for projects which would increase food production in developing countries, and would also "...improve the nutritional level of the poorest populations in the poorest food deficit countries...[and] in other developing countries." (Articles of Agreement, Article 7, Section 1, [d], [i, ii].) In its origin as an idea, IFAD was probably the brainchild of the Secretary-General of the World Food Conference---Sayed Ahmed Marei, the Egyptian Minister of Agriculture. The idea proved attractive to at least some members of OPEC, primarily because that organization was incurring sharp and increasing criticism from non-oil producing Third World countries who have been subjected to a sixfold increase in oil prices. The OECD countries, and principally the U.S., were opposed to the creation of new international organizations, but they were attracted to the idea of securing OPEC's financial contributions to aid food production meant 50-50 to the OECD members, but to OPEC equity could be defined as no more than a 60-40 ratio, with OPEC on the lesser side. Finally, by late 1977, a compromise was agreed to: a 58-42 ration, with the total fund for 1978-1981 amounting to just over one billion dollars (actually $1,021 million, with Category III nations pledging $19 million). The contribution of the U.S. was $200 million.

IFAD's financial troubles began with the first replenishment, which was to be for the 1981-1984 period. The target figure of $1.1 billion was arrived at without much difficult negotiation, with approximately the same ratios prevailing. But the U.S. arbitrarily cut its finding from $200 million to $180 million, and soon fell behind in its payments because of Congress' refusal to appropriate any funds, even in annual increments. However, other OECD countries finally agreed to fill in the deficit in the Category I pledge. On the Category II (OPEC) side, the Iranian revolution turned that nation into a non-contributor; then Libya defaulted, as did Iraq after the Iran-Iraq war began.

The second replenishment, for 1985-1987, was even more complicated. Indeed, our sketch does little justice to its complexities. A dramatic reduction in oil prices put all OPEC contributions at risk, and President Reagan and the U.S. Congress vied for honors in setting up replenishment obstacles in regard to the U.S. contribution. Tentatively, an agreement was arrived at in January 1986 for a second replenishment of just over $500 million (Category I, $300 million; Category II, $200 million; Category III, $24 million) although it seems unlikely that those amounts will actually be pledged, and if so that the pledges will be honored.

Despite this severe handicap, IFAD's record is noteworthy. In some eight years, IFAD has financed 177 projects in 87 developing countries at a total most of almost $9.1 billion, of which IFAD's share has been $2.1 billion. That is, even in the African nations some 47 percent of the project costs were incurred by the recipient government, while 28 percent were co-financed (meaning, primarily, by the World Bank and regional banks).

The politics involved in IFAD's policy cycle are quite complicated. Our description and explanation is somewhat superficial, but hopefully our generalizations will be sufficient to respond to the question of: who governs IFAD? As stated previously, the agenda of IFAD is quite closely stated in its constitutional documents; for example, <u>Lending Policies and Criteria</u> (December 1978, Article 27) specifies that projects and programs most "...normally...provide proportionately large benefits to the poorest segments of the population when compared with other groups." Because IFAD has a remarkably small bureaucracy (84 professionals and 106 support staff, as of February 1986) the matter of identification and formulation of project proposals which are in accord with IFAD purposes and criteria has been a particularly difficult matter. Through the use of technical assistance grants and special programming missions IFAD has succeeded, by and large, in influencing this process in a manner which enables the organization to function at this stage in reasonably close accord with its objectives.

It is difficult for us to generalize about the legitimation stage. The Governing council has almost no role; the Executive Board is influential in molding and modifying project proposals, but to what extent and in what matter we are uncertain. The Executive Board sessions are closed; minutes are kept; but (we are told) they are recorded only in terms of decision, not verbatim. Albeit, we do not desire to leave the impression that the IFAD President and staff govern. The Executive Board must primarily be a reacting institution, but there is good reason to believe that their reactions are critical, influential, and demanding---at least on occasion.

At the implementation stage, the limitation (in numbers) of the IFAD management staff again become apparent. Our impression, based heavily on a few interviews and a reading of IFAD's annual reports, is that the bureaucracy has been gradually gaining effective control over its implementation process. More accurately, the principal responsibility for implementation lies with the recipient Third World nation; IFAD's responsibility is "...to ensure timely and effective implementation of its projects.", and this it seems to be accomplishing. Concerning the evaluation stage the IFAD staff has been especially active and innovative, again within the confines and constraints imposed by a shortage of professional staff. Robert Berg (1986: 25) has recently commended IFAD: "...they set up monitoring authorities at the project level and strengthen evaluation at the ministerial levels [of the recipient country]." And the UN's Joint Inspection Unit report (1985: 24), noted that "...IFAD has continued to establish and strengthen monitoring and evaluation as a central element of its programme."

What does all of this mean in terms of IFAD as an organization of power? The current situation has evolved into a kind of political enigma. There is a strong ideological support for IFAD, particularly from liberals, although IFAD is not without conservative support. On the other hand, IFAD continues to face serious funding difficulties. Nearly all of the media's adverse criticism has been directed at the Reagan Administration because of its obstinancy towards the funding of IFAD, but it seems to be a fair deduction that most of the other OECD nations (and notably the major contributors to IFAD) and all the members of OPEC are permitting the U.S. to receive the opprobrium for niggardliness which they would incur, too, if put into a situation where they would be committed to increase their financial commitment to IFAD.

THE ROME FOOD AGENCIES; TOWARD 2000

We conclude this working paper with a brief foray into the hazardous field of forecasting. That is---What do we guess the organization of power in the world food agencies will look like by the year 2000? We see four possible alternatives, and each of them will be discussed in a kind of overview manner.

Unification

In a recent article, John Gerard Ruggie (1985: 353) observed that "...it was not international bureaucrats but national governments that established...no fewer than four international agencies dealing with food and agriculture alone." Political scientists have a kind of abiding passion for advising against overlap and duplication in the structure and functions of governments, and the Rome food agencies, at least deductively, seem to be a logical target for reorganization and consolidation.

For over a decade there has been an unorganized but articulate anti-World Food Council (Paris, June 1985), an "advisory group" of three was set up to investigate what the future role of the World Food Council should be. Their preliminary report has now been released (World Food Council 1986: 9), and a terse summary of their nine recommendations would be that the Council should be strengthened in its authority, funding and independence. This report will not silence WFC's detractors, but our surmise is that the end result will be a marginally stronger Council.

Then, there is the school of disbelievers concerning the IFAD. That is: Why do we need an IFAD when we already have an IDA (International Development Association---the soft-window, concessional arm of the World Bank) and three regional banks? There is no intrinsic need for a duplicative specialized agency; besides, the pie of development assistance is shrinking so a competitive structure is not only unnecessary but decisive and damaging. (Or so the anti-IFAD argument goes).

There has never been much discussion regarding the abolition of the World Food Program, at least to our knowledge, but there is concern (and notably so within the WFP) as to its increasing involvement in worldwide disaster relief. Also, and presently, there is some discussion about the future of the Committee on World Food Security, which was established by the World Food Conference to monitor and recommend regarding those concerns; it meets annually in Rome, and reports to FAO and the UN General Assembly.

There will be no unification of world food agencies by 2000; at least that is our forecast, and primarily for two reasons---one negative, one positive. On the negative side, the Food and Agriculture Organization is still suspect throughout much of the world food policy network. We feel that a considerable amount of criticism of FAO has been unfair and misdirected. Nevertheless, its image has been blemished; it is improving, but in the pejorative sense of the term the FAO continues to be viewed as a "bureaucracy"- over staffed, overpaid, unimaginative, not at the cutting edge of current agricultural science and technology.

On the positive side, the other three world food agencies (with the World Food Council as somewhat problematical) have proven themselves. There are those who are opposed to food aid, multilateral and bilateral, and their disincentive arguments constitute a valid concern. But these criticism have been heard and the necessary preventive measures factored into WFP policies. (Considerably more so, incidentally, that we believe to be true of the P.L. 480 program of the U.S.). IFAD has yet to prove that its policies and programs will enable those in rural poverty to improve their degraded condition, at least in substantial members, but there is considerable evidence that IFAD has a positive image within the world food policy network. To be sure, its funding problems are immediate and impairing, but they do not appear to be unresolvable.

Major Expansion of Authority

This is an unlikely alternative, in our view. Neither the World Food Council nor FAO will be permitted by the OECD, or even by the Group of 77, to become a World Food Authority. Lord Boyd-Orr's dream continues, but in that form. The World Food Program will not be transformed into the world disaster agency; there are too many other international disaster agencies who would engage themselves with ardor and vehemence in that kind of turf fight. Besides, this would require a considerable reconstitution of the WFP as an institution; there is more to disaster relief than the supply of food, important as that is. Likewise, the World Bank's IDA is not going to be dissolved into an IFAD, although a considerable increase in oil prices would renew the appetites of OECD members toward finding ways to direct substantial amounts of those "undeserving" profits into development assistance. In this kind of thinking, IFAD is not without growth potential, although one is surely pressed to think of ways that would convince OPEC that the unpleasant is the necessary.

Dissolution

Cynics (or are they realists?) often conjecture that international bureaucracies, like those of the national variety, are rarely abolished and seldom tend to fade away. There are possibilities here, although we see the probabilities as quite high that not one of these world food agencies will be abolished. Our reasons can fairly well be extracted from the prior discussion concerning unification. If the World Food Council were to be abolished, who would assume its persuading, mobilizing, coordinating responsibilities? The FAO would be the logical, but politically unviable choice. In the best of all possible worlds, food aid should become an anachronism, institutionally and policy wise, but that is a 21st Century dream. Besides, the WFP is politically useful as a multilateral agency, performing functions that OECD surplus-producing nations would find to be awkward, perhaps embarrassing, and possibly counterproductive, if they had to be carried out exclusively within a bilateral framework. IFAD is a candidate for dissolution, but not realistically; this may not be the heyday period for liberals, but we see IFAD as institutionally invulnerable, although its funding arrangements will continue to be the subject of hard negotiating.

Incremental Change

For reasons we do not need to examine, political scientists are generally (perhaps genetically!) incrementalists. Things do and should change, but not very dramatically, and very likely in the direction of more redistributive policies. Such, at least, is the conventional wisdom in political science. In this kind of philosophical-analytical perspective we view world food organizations, too. There will necessarily be policy changes, but we do not see them to be of a dramatic nature. Likewise, their decision-making processes have not changed much in the last decade. At its 1986 session in Rome, the World Food Council reverted, at least in part, to the format and agenda for which it was originally designed; i.e., a policy-innovation kind of arrangement orchestrated by the Council's Executive Director and his staff, at which ministers of agriculture, and a few of their money- and planning-oriented counterparts, would talk over what they are, might be, and could be doing. And the final report constituted a kind of "sense of the meeting", experiential in content, rather than a much politicized set of conclusions and recommendations. Whether these high-level, time-conscious political administrators will accept this format as both utilization and perception-broadening, and not just another cacophony of rhetoric, remains to be seen.

In this increasingly interdependent world there is a vital need for a Food and Agriculture Organization. And in several policy areas---food standards, plant and animal genetics, seed identification and preservation, the use and sale of pesticides, fungicides and rodenticides, among others---we believe that FAO's power and responsibility will be gradually increased.

The World Food Program is now adjusting its recent increase in authority, which were granted following the deliberations of the Joint UN/FAO Task Force on WFP Relationship Problems. WFP management "won" more autonomy from FAO in internal matters involving personnel, financial and audition. But that power struggle seems now to have been resolved, and WFP still does not have the status of an independent specialized agency. However, WFP will likely have incremental increases in its authority over the next few years---funding of programs (not only projects), multi-year funding of projects and programs, as examples.

Just how the funding difficulties of IFAD will be resolved (meaning alleviated; funding is never finally resolved) is unclear to us. The two principal actors in this matter seem to be the U.S. and Saudi Arabia. If so, one could develop a few scenarios sketching out how this accommodation might come about, but that is not our task. In our judgement, the U.S.---in its own interests, and, mutually, those in Third World nations---should retreat from two wrong-headed positions. Namely, that funding for IFAD must be niggardly and defensive, and that IFAD's staff needs are now being adequately served. We have no grandiose numbers in mind; perhaps a return of the original one plus billion dollars for a three-year period and a doubling of the management staff. But IFAD has earned its right to be treated with respect, and adequate funding should be subsumed therefrom; that right should be acknowledged and then accorded its just deserts.

To conclude, our own intellectual perspective comes out of a kind of pluralistic-Madisonian tradition: "Men [and women!] are not angels and angels do not govern men"---"the latent causes of faction are ... shown in the nature of man"---"ambition must be made to counteract ambition"---"... experience has taught mankind the necessity of auxiliary precautions" (Federalist Papers, 10 and 15). Within this kind of philosophical outlook, we view the international food organizations as useful and, to a considerable extent necessary instruments to be used in the pursuit of justice and equity, both for rich as well as poor nations. By and large, these organizations have functioned fairly effectively, with some positive results. We view the four-organization structure as optional in terms of balancing the interests of the competing factions, while at the same time facilitating some constructive action. How they are governed, by whom---for whom---in what manner---with what results, are political matters of enduring consequence.

REFERENCES

Berg, Robert J. (1986). "Donor Evaluations: What Is and What Could Be." Paper presented at the International Conference on the Role of Evaluation in National Agricultural Research Systems, Singapore 7-9, July.

International Fund for Agricultural Development (1985). Annual Report 1984. Rome: IFAD.

_____ (1979). Agreement Establishing the International Fund for Agricultural Development. Rome: IFAD.

_____ (1978). Lending Policies and Criteria. Rome: IFAD.

Keohane, Robert O. (1984). <u>After Hegemony: Cooperation and Discard in the World Economy</u>. Princeton: Princeton University Press.

Keohane, Robert O. and Joseph S. Nye (1977). <u>Power and Interdependence: Politics in Transition</u>. Boston: Little, Brown and Co.

Kriesberg, Martin (1984). <u>International Organizations and Agricultural Development</u>. Washington, D.C.: USDA.

Moyer, Wayne (1986). "FAO as a Structure of Power: The reality of Its Limitations." Paper presented at the annual meeting of the Midwest Political Science Association, April.

Phillips, Ralph W. (1981). <u>FAO, Its Origins, Formation and Evolution 1945-1981</u>. Rome: FAO.

Ruggie, John Gerard (1985). "The United States and the United Nations: Toward a New Realism." <u>International Organization</u> 39, 2 (Spring).

U.S. Senate. Select Committee on Nutrition and Human Needs Staff Report (1976). <u>The United States, FAO and World Food Politics: U.S. Retaliations With an International Food Organization</u>. Washington, D.C.: US Government Printing Office.

World Food Council (1986). <u>Recommendations and Suggestions For the Future</u>. WFC/5.

World Food Programme (1978). <u>World Food Programme Basic Documents</u> 4th ed. Rome: WFP.

12

U.S. Foreign Agricultural Policy and the Less Developed Countries

Jean Doyle

INTRODUCTION

United States foreign agricultural policy since World War II has had a largely negative impact on agricultural development and food consumption in the less-developed countries (LDCs). Although this result has not necessarily been a conscious, premeditated objective of that policy, it is a perhaps unavoidable by-product of an American foreign policy based on the premise of U.S. strategic, economic, and political dominance. The fact that both foreign agricultural policy and policy toward the LDCs are derivative of and subordinate to American foreign policy as a whole, means that direct cause and affect relationships between the two are difficult to pinpoint. Nevertheless, on balance, the LDCs have been more victims than beneficiaries of American foreign agricultural policy during the post-war period.

An accurate analysis of how and why U.S. foreign agricultural policy has adversely affected the LDCs requires an understanding of the policy process in this area. That process has been characterized by shifting sets of actors and relationships among them, changing policy objectives in response to events, and a belated awareness of the part of U.S. policy makers of the consequences of their decisions on the LDCs. Although a chronological examination of these factors can help us to understand trends and motives, it fails to capture the cumulative effect of American policy abroad. Individual policies or decisions have not, by themselves, created the food/agriculture crisis in the LDCs, but collectively and over time the combination of poorly conceived development theories, self-serving aid policies, and an emphasis on short-term foreign policy gains have had this result. By the time policy makers recognized the severity of the problem, its magnitude had overwhelmed the resources available to solve it.

ORIGINS OF THE U.S. FOREIGN AGRICULTURAL POLICY

The domestic political forces which shaped U.S. foreign agricultural policy after World War II emerged during the interwar years. As with other areas of the economy, American agriculture was profoundly affected by the Depression and by New Deal policies which were designed to prevent radical economic fluctuations through greater governmental regulation of economic activity. Although farmer interests had always had an impact on American politics, particularly congressional politics, New Deal form policies helped to expand the political influence on the farm bloc and facilitated its entrenchment in the Executive branch bureaucracy. Due to global Depression, neo-mercantilism, and American isolationism---all of which inhibited international trade---foreign agricultural policy itself was not an issue in the interwar period. However, by the time it reemerged as an issue in the wake of World War II, the farm bloc had amassed the power to determine its parameters and priorities for the next two decades.

The central problem of American agriculture during the 20th century has been overproduction. While that has provided relatively low, stable food prices to the consumer, it has also tended to depress farmer income and has led to the creation of vast commodity surpluses. Soil fertility, large quantities of arable land, agricultural research and education, and improved inputs and machinery all contributed to high levels of agricultural productivity, while demand for agricultural products remained static or declined. Overproduction, in turn, hurt the farmer whose income fell as he produced more and for whom price instability made long-term planning next to impossible. Consequently American farm policy developed around two central---and not necessarily compatible---objectives: (1) the stabilization of farm income, and (2) the elimination of commodity surpluses. It was in the attempt to implement these goals that the farm bloc acquired its decisive and enduring influence over American agricultural policy, and the government itself became a major actor in the overall policy process.

Through the 1933 Agricultural Adjustment Act and its successors, the government instituted price support or parity payments to farmers which guaranteed them at least minimum prices for their crops as a means of stabilizing farm income, in return, farmers agreed to production control such as marketing quotas and acreage allotments, and to "land banking", ostensibly for soil conservation purposes, in an attempt to curtail production. In addition, the Community Credit Corporation (CCC), established by executive order in 1933, was empowered in 1935 to make "non-recourse" loans to farmers---i.e., to buy and store surplus agricultural stocks---thereby involving the government directly in commodity markets for the first time.

In both the long and short run this policy mix failed to solve the overproduction and surplus problems. Marketing quotas led farmers to use land for alternative crops, with subsequent oversupply in these substitute commodities; acreage allotments encouraged growers to farm existing land more intensively with the result of higher per acre yields. The latter effect was particularly noticeable during the 1950s when the chemical revolution boosted per acre yields significantly (Allan, 1957). Furthermore price supports led to artificially high domestic prices for farm products, making American markets attractive to foreign exporters, reinforcing farmer predispositions toward protective tariffs, and making American commodities abroad without export subsidies. Although the 1930s trend toward economic nationalism temporarily made the export problem the contradictions between purity---induced high commodity prices and the desire for greater global trade were apparent to the Roosevelt administration and became a sticking point in the GATT negotiations after World War II (Hickman, 1949). However, once price supports were instituted as a cornerstone of U.S. farm policy they proved impossible to eliminate; American trade negotiators were forced to accept them as a given rather than being able to use their elimination as a bargaining chip to induce reciprocal concessions from trading partners. As a result, export subsidies of one form or another became a permanent feature of American foreign agricultural policy, as did tariff protection for domestic agricultural markets.

Politically, the New Deal farm program enabled farm interests to acquire decisive influence in the making of agricultural policy and both the Congressional and bureaucratic levels. Malapportionment of the House of Representatives and equal representation in the Senate traditionally had given farm interests considerable strength in Congress, and Executive branch support for government intervention in the agricultural sector helped to enhance that influence. The U.S. Department of Agriculture (USDA) was the major bureaucratic beneficiary of the New Deal programs because of its expanded regulatory responsibilities, including the establishment of market quotas and soil bank requirements and the administration of the price support program. USDA quickly came to be perceived as vital to the economic survival of the farmer, and in the pattern typical of regulatory agencies, it soon became the captive if its own clientele (Peterson, 1979: 13). Thus, the policy changes of the 1930s not only gave farmers some degree of the income stability they had sought, but it also created for them political influence through multiple points of access to the policy process and a strong bureaucratic advocate for their views in the USDA.

The Second World War dramatically and fundamentally transformed the international context of American foreign policy making. Autarky and isolationism were superceded by the Cold War and an activist international based on American strategic and economic dominance Anti-communism produced the ideological rationale and American power the means for the creation of an international system designed to serve and maximize U.S. global interests and influence.

One of the cornerstones of that system was a "free world" international economic order built around the concept of economic interdependence. This would not only further American interests, but would also deter "communist expansionism" by limiting member countries' economic options. Or, as it was more optimistically and less chauvinistically sold to potential participants, the benefits of membership would be so attractive---in terms of expanded aid, trade and access to the resources of the newly created international economic institutions such as IBRD and the IMF---that nations would choose voluntarily to participate in the new order. The Marshall Plan was touted as a working model of the mutual benefits of interdependence, obscuring the fact that it was accompanied by dollar supremacy, and "Open Door" for American goods and capital, and a freer trade in which the undamaged U.S. economy had decisive competitive advantages.

For foreign agricultural policy the key element of the new international framework was unprecedented access to foreign markets for American agricultural products. This was facilitated by (1) the transformation of the dollar into the primary currency of international trade, (2) the 1947 GATT negotiations for across-the-board trade liberalization, and (3) American insistence that the Europeans renounce their exclusive trading rights with their colonies as a quid pro quo for Marshall Plan aid. The domestic farm lobby's persistent resistance to the kind of phased, reciprocal tariff reductions envisioned by GATT was partly overcome by President Truman's executive order providing "escape clauses" (Hickman, 1949: 40-41) in future agreements for competing agricultural products and partly be a section of the 1946 Agricultural Act re-endorsing export subsidies for commercial commodity sales (Allan, 1957: 37). Despite valid objections from the State Department, which was concerned about the negative impact these measures would have on U.S. relations with friendly nations that were also commodity exporters, and charges from other countries that export subsidies and escape clauses ran contrary to the basic purposes of GATT, no softening of these farm bloc-backed provisions occurred. American preeminence was such that allies were left with no alternative but to accept the U.S. position regardless of its unpalatability.

European reconstruction and the Korean War both stimulated demand for American agricultural exports and masked the permanence of the overproduction problem until the Eisenhower administration (Rau, 1957: 81; VanGigch, 1972: 168). By then finding storage capacity for government-held grain surpluses had become a serious problem and the costs to the government (and ultimately to the taxpayer) of absorbing excess commodities were getting out of hand. This not only made the government a major actor in commodity transactions, but it also provided a stimulus for more active government efforts to find ways of disposing of the surpluses. It was a formiddle task, one which had to satisfy a variety of competing interests to have any hope for success.

PL 480, the Agricultural Trade Development and Assistance Act of 1954, was the final compromise which emerged and shaped foreign agricultural policy for the next two decades. The bill's title was an accurate reflection of its priorities. Title I provided for "the sale of U.S. surplus agricultural commodities to friendly nations with payment in the currency of the recipient nation" (Tweeten, 1979: 453), i.e., outside the dollar market where such sales might compete with normal commercial transactions (Yost, 1971). For this reason it represented a clearcut victory for the farm bloc. Three assumptions underlay the concept of foreign currency sales. First, it was believed that real demand for U.S. agricultural products did exist on a global basis (the notion of additionality), but that may countries were unable to buy commodities in commercial markets due to balance of payments difficulties or temporary shortages of covertible currency (dollars). For countries in this situation, Title I sales would help to see them through their short-term problems while simultaneously reducing U.S. government surpluses and earning money to pay for overseas government operations. Second, it was assumed that the lack of commercial demand was temporary; therefore the establishment of American sources of supply assisted by vigorous market development efforts would eventually lead to permanent commercial markets for U.S. commodities. Finally, policy makers still viewed overproduction, and thus the need for Title I, as transitory phenomena which would disappear as production and global demand achieved equilibrium. All of these assumptions proved to be faulty, but the fact that they were believed in 1954 enabled both government and farm interests initially to support the bill.

What loosely might be called the humanitarian lobby also favored passage of PL 480 because of its Title II provisions for disaster and emergency relief. Although these groups were "only marginally influential" (Peterson, 1979: 68) in the legislative and administrative arena, their support helped to broaden the base of the coalition favoring the bill. Moreover, the inclusion of humanitarian goals was good public relations: the veneer of concern for the world's less fortunate helped to sell the bill to the American public and lent a sense of higher moral purpose to what was essentially a self-interest and self-serving piece of legislation.

Foreign policy concerns were clearly subordinate to surplus disposal during the assuage of PL 480 and the State Department was initially unenthusiastic about the bill. There were fears that it would "(1) alienate allies of the United States who were also food exporters, (2) decrease the emphasis in the new nations on improving their own food production, and (3) foster an attitude in Congress which favored more support for PL 480 at the expense of the dollar aid program" (Talbot, 1968: 305). These concerns were well-founded. Title I sales did constitute a form of dumping in international markets and Title II government-to-government conveyance was clearly state trading both of which ran counter to the spirit of GATT and engendered the hostility of friendly foreign competitors.

The State Department correctly anticipated Congressional reactions for PL 480's impact on other foreign aid programs. Given both the continued availability of government commodity stocks as a foreign assistance resource and the absence of a strong domestic constituency for foreign aid in general within the U.S., Congress did ultimately force the

State Department to substitute PL 480 aid for dwindling dollar aid programs (Saylor, 1977: 203).

It also indicates that from the inception of PL 480 there was a concern within the State Department as well as in the humanitarian community about the bill's effect on agricultural development in recipient countries. Again this concern was prophetic if ineffective at the time it was voiced by the handful of development economists who recognized the disincentive effects of food aid on developing nations' food production. But the development economists were a small, highly specialized group struggling for recognition within their own discipline and a consensus within their subspeciality. Their theoretical projections were overpowered by the irresistible attractions of immediate surplus disposal, and at least in the short run, the economists adapted their theories to the prevailing political reality.

The contradictory goals of PL 480---trade development, surplus disposal, humanitarian relief, foreign policy tools---were resolved in the early years of implementation in favor of surplus disposal, as Peterson has ably documented. Despite the multitude of government departments, agencies, bureaus, and committees with a stake in different aspects of PL 480, the USDA dominated the initial struggle between itself and the State Department over control of implementation. Agriculture officials and farm bloc Senators repeatedly accused the State Department of hamstringing or slowing down. Title I sales by emphasizing the foreign policy implications of proposed sales agreements. Furthermore agricultural interests, fearful that foreign currencies accumulated through the Title I programs might be sued for agricultural development projects abroad, tacked on the so-called Cooley load amendments in 1957 which prohibited the use of these funds for such purposes (Peterson, 1979: 58-63).

This fear of losing markets as a result of aid-resisted development program characterized the entire first phase of postwar foreign agricultural policy, effectively limiting the types of aid that the State Department could offer other countries and disregarding consideration of whether American interests were served or harmed by such a policy. Only when the "communist menace" could be invoked to justify specific sales proposals was State Department approval virtually guaranteed. Title I sales to Burma, Egypt, India, Taiwan, and Korea in 1956 were designed to counter perceived Soviet attempts to "subvert and communize" these nations and PL 480 aid to Yugoslavia was a transparent attempt to wean Tito from economic dependence on the Soviet bloc. When foreign policy and farm bloc goals for PL 480 coincided, therefore, the interagency rivalries were subordinated to shared mutual interests (Peterson, 1979: 63).

Although PL 480 was designed primarily to help solve the domestic overproduction in the U.S., its passage also put in place an important element of American policy toward the LDCs. LDCs---except those defined as being threatened by communist subversion---were not a central focus of U.S. foreign policy during this period. However, policy makers did recognize that the promise of rapid economic development was the carrot which would entice LDCs to participate voluntarily in an interdependent, pro-capitalist global economic system and they endorsed a policy matrix designed to achieve this goal. That paradigm, based on the success of the Marshall Plan in revitalizing the European economy is less than a decade, assumed that large infusions of aid, investment and expertise could produce accelerated development in less developed economies as well. W.W. Rostow (1960) best reflects this approach. Aid, investment, and expertise provided by "developed" nations and channeled into programs of rapid industrial expansion would serve as a catalyst for eventual self-sustaining economic growth in the LDCs. In this model, modernization was equated with industrialization, whose benefits would "trickle down" to other economic sectors over time, pulling them along into the modern world.

This industry-led development paradigm assumed that conditions in the postwar global economy made it possible to substitute aid and investment for the prolonged process of capital accumulation that had occurred in Western states. Agriculture was viewed as the most backward, least productive economic sector; therefore investment in agriculture was considered a squandering of scarce resources and lacked to the projected spillover benefits of investment in industry. Furthermore, the availability of large commodity surpluses in the U.S. through programs such as PL 480 meant that food imports could be sued to fill any supply gaps which might emerge before agricultural modernization began. Thus since agriculture was not needed for capital accumulation, since investment in it was unproductive relative to other uses, and since food self-sufficiency was deemed unnecessary, the agricultural sector could virtually be ignored in the early phases of the development and modernization process.

As was true of PL 480, the assumption behind this hypothesized model of development proved to be flawed. The

Marshall Plan analogy simply ignored the differences in culture, resources, land tenure patterns, Western-educated LDC elites share capitalist goals and behaviors, and that nationalism was the same as national unity. It was assumed that adequate quantities of affordable food imports would continue to be available indefinitely. And perhaps most importantly, it failed to comprehend that economic "take off" required the participation of the rural population, but that the neglect of agriculture made such participation impossible. However the problems, including dependence, massive indebtedness, and distorted development, were largely masked by the apparent early successes of the model which seemed to be benefitting both donor and recipient nations.

By the end of the 1950s American foreign agricultural policy, primarily through the vehicle of PL 480, had become a means of disposing of domestic agricultural surpluses abroad on bargain basement terms. Farm bloc control of the agricultural policy making process produced this outcome and it prevented U.S. foreign aid from being used to foster agricultural development abroad for fear of losing markets for American commodities. Economic development theory (perhaps unwittingly, perhaps pragmatically, but certainly conveniently) accommodated this position by dictating that rural and agricultural development would occur as a result of rather than as a prerequisite to industrialization. In the interim, potential shortfalls in the food supplies of the LDCs could be filled by imports, more profitable export-oriented cash crops could replace food crops to bolster LDC foreign exchange earnings, and the rapidly growing urban areas at least temporarily would rely almost entirely on imported food. The disastrous results of this formula were not understood until it was too late.

The 1960s represented a transitional period for U.S. foreign agricultural policy, one in which international economic concerns increasingly supplanted domestic considerations in the policy-making process. During this decade, a combination of circumstances---the relative weakening of American's global economic position, greater foreign policy emphasis on less-developed countries, and declining farm bloc influence in domestic politics---resulted in the downgrading of surplus disposal as a foreign agricultural policy goal and to the increasing coordination of agricultural programs with broader international economic objectives. Foreign agricultural policy became more subtle and complex as policy-makers began to recognize and to manipulate American's comparative advantage in commodity production, gradually transforming what had been viewed as a curse into a foreign policy weapon.

On the international front, the U.S. found itself forced to adjust to an altered environment in the Sixties. European and Japanese economic recovery led these areas to begin competing with the U.S. for markets, outlets for capital, and raw material supplies. Dollar dominance became a dollar glut, especially during the second half of the decade with America's military ventures in Southeast Asia. Consequently the U.S. began to experience difficulties with its trade and payments balances, a situation which could not be ignored without jeopardizing its overall global economic interests.

Simultaneously, the focus of U.S. foreign policy was shifting from Europe to the Third World. Economists were acquiring a more sophisticated understanding of the economic development process and were frightened by the data being generated about the LDCs. The statistics projected rapid rates of population growth and a widening food-population gap which augured ill for the political stability American experts deemed vital to the U.S. interests in these areas.

At home, demographic change had also occurred, with rural out-migration causing the farm population to drop by more than one-fourth by 1970 (VanGigch, 1972: 170). This depleted the voting strength of rural constituencies, weakening the farm bloc's ability to influence foreign agricultural policy and making it easier for urban-based Democratic administrations to reorient that policy toward American global economic goals. Although the agricultural lobby continued to exercise decisive control over domestic farm policy during the 1960s, its relative electoral and bureaucratic positions were slipping. Other interests---the foreign policy establishment, development economists, agribussinessmen, agricultural input suppliers---were developing a growing stake in foreign agricultural policy and they began to compete successfully with the entrenched farm lobby for a voice in the shaping of policy.

The reorientation of foreign agricultural policy and the adoption of a new paradigm for economic development in the LDCs formed the response to these pressures and changes. Basically, the new economic development paradigm repudiated the previous assumption that economic development---and particularly agricultural development---abroad were inherently contrary to American interests, i.e., that it would eliminate markets for U.S. exports. Agricultural development was not to be sanctioned and even underwritten by foreign aid because over the long term the U.S. could not continue to feed the world (President's Report, 1966). both foreign policy makers and LDC leaders understood that the U.S. had consciously pursued a policy of encouraging food dependence for a decade and that it would be held responsible if adequate food supplies were to disappear. At a time when winning hearts and minds in the LDCs was in the front line of the Cold Ward conflict and maintaining political stability abroad was a prerequisite for profitable overseas American investment, the prospect of hunger-induced chaos in the Third world was both alarming and counter-productive.

Implementation of the new paradigm required adjustments to PL 480 as well as related policy initiatives. The 1966 Food for Peace amendments dropped the Cooley loan restrictions on foreign development projects; Title I sales and other forms of foreign assistance were now made contingent upon recipient countries placing "an appropriate emphasis on agricultural development and related projects." (USDA, 1967). Although enforcement of these self-help clauses was weak (or totally absent in the case of PL 480 sales to Vietnam and Korea in the late 1960's), the U.S. was at least giving lip service to the self-help concept. In addition, American aid programs required recipients to engage in more vigorous voluntary efforts at population control (President's Report, 1966). The birth control provisions were unpopular and brought forth accusations of racism, genocide, hypocrisy, and unwarranted violations of sovereignty, but they were nonetheless included in aid agreements and monitored with some degree of rigor.

A second 1966 amendment to PL 480 was designed both to relieve the pressure on the deteriorating U.S. balance of payments position and to encourage greater LDC food self-sufficiency, in that order. It required that Title I foreign currency sales, called soft sales, be phased out entirely by 1971 and that dollar (hard) sales, subsidized by long-term credits, be substituted in their place. In other words concessional sales were to be continues as a form of aid, but on terms more favorable to the U.S. The primary reason for the change was the during twelve years of soft sales the U.S. had accumulated vast holdings of certain foreign currencies which were far in excess of American in-country needs. For example, massive Title I sales to India had given the U.S. control of close to one-quarter of the Indian money supply, a situation which threatened to undermine that country's monetary policy and seriously compromised its fiscal sovereignty. Dollar sales, on the other hand, even with credit and export subsidies, would be eventual money earners for the U.S. and would help to offset the growing deficits in its non-agricultural trade balance.

Secondarily, it was hoped that by increasing the costs of food imports in this manner, LDCs would be prodded into giving greater attention to agricultural development. Unfortunately, however, this did not happen, and the rising costs of food imports contributed to the balance of payments crisis in the LDCs. Even if the political and material resources for crash programs of agricultural development had existed---which they did not---five years was simply to short a time to reverse the effects of two decades of food dependence and agricultural sector neglect. If the official data are correct, the hard currency amendment added at least $300 million to the outstanding debt of recipient countries by 1971[1]. Since the balance of payments game is zero-sum, this helped the U.S. by hurting those who could least afford it.

IMF loan policies that were supposed to ease the debt burden of the LDCs compounded the problem of food import substitution. The IMF's balanced budget requirements reduced the quantity of resources available for development programs in general and therefore the likelihood of funding for rural development programs. Its demands for currency revaluation made both food imports and the debt servicing to pay for them more costly. However rather than encouraging investment in food production as a means of paying higher import bills. Cash crop production was also the more likely magnet for foreign investment due to its higher profitability and its greater suitability for industrial-style organization and management. Thus IMF policies, while perhaps not intending to do so, helped to inhibit the LDCs' ability to expand food production as a means of reducing their foreign debts.

The notion of encouraging greater LDC food self-sufficiency also found considerable support from American agricultural input suppliers who had discovered by the 1960s that domestic demand had peaked and they were beginning to seek international markets for their products. With funding from the Ford and Rockefeller Foundations (which perhaps only coincidentally were "related" financially to Ford Motor and Standard Oil, two of the largest input suppliers), plant specialists had undertaken a major effort to adapt high yield, scientific farming practices to tropical climate agriculture. What came to be known as the Green Revolution was eagerly adopted by U.S. policy makers and was promoted through American aid programs to countries like India and Pakistan---who were among the largest food aid recipients.

For the policy makers, the Green Revolution had the dual attractions of offering a quick fix solution to the problem of food deficits and a longer term boost to the ailing American balance of payments position. The latter was achieved primarily through aid-typing, a practice which required aid recipients to spend aid dollars on purchases from the donor country. Green Revolution inputs such as fertilizers and pesticides were particularly beneficial to the donor of tied aid because they were non-durable goods and required replenishment for each agricultural cycle. And although Green Revolution agriculture did hold out the promise of higher yields and greater food self-sufficiency to the LDCs, it involved tradeoffs---higher costs, increased risks, and a high level of energy intensiveness.

The new emphasis in American policy, however, tended to ignore the cumulative, contradictory effects of previous development decisions. The number of new jobs created by industry-led planning in the LDCs had failed to keep pace with the rapid population growth resulting from improved public health work and upgraded infrastructure. Despite the adoption of Green Revolution technology, LDC agriculture was unable to feed a larger population because of (1) the extensive substitution of export (foreign exchange earning) crop production for food crop production, (2) the lack of structural change in the food-producing segment of the rural economy (3) the existence of probing and fiscal policies which served as disincentives to food production, and (4) the absence of investment in food production accompanied by resistance to such investment on the part of LDC elites who had been encouraged by the development economists to believe that it was both unnecessary and unproductive.

Thus by the mid-1960s, when Western development planners were trying to reorient development strategy toward agricultural modernization, the conditions in the LDCs militated against such an effort. Agricultural sector neglect and population growth were jeopardizing LDC capabilities for economic modernization and participation in the international economic system as contributors rather than as perpetual debtors, yet the changes in U.S. foreign agricultural policy to exacerbate rather than to solve the fundamental problems. America's brief flirtation with basic structural change (i.e., land reform) in the agrarian sector through the Alliance for Progress was aborted when its potentially destabilizing effects came to be understood (Crawford, 1972) The U.S. requirement that vital food imports, even concessional food aid, the paid exchange resources. These mixed signals, unsupported by adequate levels of dollar aid for agricultural development, led to confusion in the LDCs and to growing disillusionment on the part of their leaders with the efficiency of Western development planning.

The opening years of the 1970s were characterized by a "business as usual" attitude in the U.S. toward foreign agricultural policy, based on preexisting assumptions about global agricultural conditions; that the U.S. continued to have considerable excess production capacity; that the rate of yield-expanding technological and biological innovation would remain high; that the LDCs capacity for food production was growing to the dissemination of the Green Revolution; that the food-population rate was being won. American commodity exports were growing at a steady, if unspectacular pace, and surplus production still remained the central problem of domestic farm policy. Since the Malthusian' predictions about catastrophic food shortages had failed to materialize, the U.S. could afford to take a bold new foreign policy initiative without jeopardizing either its own international interests or the lives and health of the world's neediest people.

That initiative took the form of the 1972 Russian grain sales, Henry Kissinger's carrot to the Soviets to whet their appetite for detente. 1972 was also an election year; a large Soviet-American commodity transaction would score points with the traditionally Republican farm bloc and help compensate for rising anti-Administration sentiment generated by the unpopular Vietnam War.

Unlike previous government-sponsored commodity transactions, which had been limited to the disposal of surplus stocks, the 1972 agreement authorized Soviet buyers to deal directly with American grain trading companies without government intervention. Therein lay the source of the impending food crisis, as Destler (1980: Ch.3) has ably documented. Once the Russians were given the go-ahead to enter commercial markets, they did so with a vengeance, buying 433 million bushels of wheat, 6.5 million tons of feed grains, and one million tons of soybeans during the months of July and August. These purchases represented "half of the U.S. carryover stocks [of wheat] on July 1, 1972, and a bit over a quarter of 1972 production" (Destler, 1900: 35). Since they were cash transaction with private corporations, the government had no mechanism in place at the time for the monitoring of purchases. Both the method and magnitude of the sales were unprecedented: within two months, American grain reserves had become seriously depleted and a substantial portion of the standing crop was committed for export, all in the absence of government knowledge or control. The results---inadequate buffer stocks and upward pressure on domestic food prices---set the stage for the subsequent world food crisis and conflicts over U.S. foreign agricultural policy.

Under what had come to be considered normal conditions, the impact of the 1972 grain sales would have been a minor irritant to American consumers and a long-awaited windfall for agricultural producers. However, worldwide economic conditions were far from normal in 1972-1973. Oil-producing countries, newly united and conscious of the industrialized nations' dependence on their product, quadrupled the price of oil in less than a year. For big oil importers like the U.S., the OPEC move wreaked havoc with the balance of payments and fueled both direct and indirect rises in consumer prices. Retail food costs jumped sharply, for agriculture had become on of the country's most energy-intensive industries. The Green Revolution in the LDCs faltered as pesticides, fertilizers, and irrigation pump fuel became unaffordable or unavailable. Food production there declined when "many farmers reverted to less productive, cheaper methods of farming (Simon, 1979: 9).

The weather in many parts of the world for the 1972-1973 growing season compounded the problem. Drought in the Sahel region and on the Indian subcontinent created, respectively, devastating famine and the return of serious food deficits. Requests to the U.S. for disaster relief poured in, further stimulating demand for limited commodity supplies. At the same time, the normal commercial demand for feed grains from Europe and Japan remained strong, and because these were the hard currency transactions most beneficial to the U.S. balance of payments position, the buyers were in a strong competitive position to bid for available supplies.

The end result of this combination of factors was a fundamental transformation of the world food situation from one of abundance to one of scarcity. For American policy-makers, old assumptions and priorities were no longer valid as their task shifted from that of finding ways of relieving the food glut to that of allocating scarce food resources among competing interests and objectives. Foreign agricultural policy became a subset of a more general food policy in which achieving one set of goals meant sacrificing other. With scarcity, the contradictions between farmers, consumers, foreign policy-makers, humanitarians, and food-deficit LDCs became irreconcilable, no longer simply differences over means but conflicts over the outcome of the allocation process itself. After a number of false starts, a reasonable coherent foreign agricultural policy did emerge, but it was one that predictable left the food-dependent LDCs at the bottom of the priority list.

CHANGES IN THE U.S. FOREIGN AGRICULTURAL POLICY SINCE THE 1973 WORLD FOOD CRISIS

In the domestic political arena, the battle over food policy initially saw consumers, organized labor, the farm bloc, and the economic policy community pitted against each other for control of that policy. American consumers had become accustomed to low, stable food prices and were outraged when the Soviet grain sales triggered rapid food price inflation. Why, consumers asked each other and their elected representatives, should their standard of living be compromised so that the Russians---of all people---could be better fed? Shoppers began to boycott high-priced meat, and they sent the message to Congress that they, as voters, expected their economic interests to take precedence over potential foreign policy gains in the policy-making process. President Nixon acknowledged consumer concern as he inched toward the 1973 soybean embargo by declaring that in allocating the products of America's farms between markets abroad and those in the U.S., we must put the American consumer first (Destler, 1980: 59fn23).

The consumer had a political ally in the economic policy community: the Council of Economic Advisers, the Treasury Department, and Cost of LIving council, and the Federal Reserve Board (Destler, 1980: 46-47). Its concern transcended specific pocketbook issues, centering instead on the overall strength of the economy and its effects on America's global economic position. Although this group's influence grew as the need for greater food policy coordination became evident, it initially had failed to foresee the ramifications of the Russian grain sales and thus had offered only the weakest resistance to them.

Farm bloc attitudes toward the Soviet purchases were ambivalent. On the one hand agribusiness---Agriculture Secretary Earl Butz's primary constituency---was delighted with he principle of sustained access to untapped Eastern bloc markets. Small-to-average producers, however, were unhappy with the government's faulty handling of the accord's implementation in which the bulk of the profits had accrued to the giant grain corporation such as Continental, Cook, and Cargill rather than to the producers themselves, Farming interests were united, however, in their resistance to Secretary Butz's proposal to abolish established government farm programs and let the free market prevail now that overproduction had ceased to be a problem (Destler, 1980: 17). Price supports and acreage restrictions, although unused during the period of greatest demand (i.e., through 1977), remained in effect as a safety net for farmer income, as did the quota system for non-grain products such as sugar, peanuts and tobacco.

The farm bloc prevailed on the issue of farm programs, but it was thwarted in its bid to prevent the imposition of export controls on agricultural commodities. Both the soybean embargo of 1973 and the implementation of export controls in 1975 on Soviet grain sales represented setbacks for producers and were indicators of the farm lobby's declining influence in the policy-making process. Consequently, farmers were having to consider compromise and coalition-building as a means of generating support for beneficial policy outcomes.

Consumers were an unlikely ally, since they had cost themselves as the victims of the food policy drama. Although consumers were concerned primarily with price inflation, their movement was also becoming involved in health, nutrition, and environmental issues related to food and food policy, to the irritation of both farmers and processors[2]. Organized labor was also an improbable coalition partner. It had perenially been at loggerheads with agribusiness over the unionization of migrant labor, its members were hard-hit by food price inflation, and it had actively assisted the consumer movement with George Meany's 1975 decision. As Destler (1980: 105) explains "that members of the International Longshoremen's Association would refuse to load grain bound for Russia unless the Administration promised that the interest of consumers---and the shipping industry---would be protected." The latter move proved crucial to President Ford's decision to impose export controls in 1975.

However, the interests and circumstances often combine to create strange bedfellows, as the general domain of food policy-making was about to prove. The denouement came in 1977 with the passage of the Food and Agriculture Act. As Don Paarlberg summarized the situation, "what had been thought to be a shootout developed into a love feast." (Destler, 1980: 53). Considerable logrolling by all groups-indicative of their flexibility an willingness to bargain-facilitated the process, allowing agribusiness, consumers, and labor all to claim victory to their respective constituencies. Farmers salvaged their commodity programs by agreeing to substitute deficiency payments based on costs of production for the outmoded parity formulae, and they accepted ceilings ($45,000 in 1977) on the amount of annual payment any single farmer could receive from the government. Both of these provisions mollified consumers who preferred to pay the cost of farm programs as taxpayers rather than at the grocery store and were satisfied that finite limits now existed for government "handouts" to commodity producers. In return, the farm bloc supported the food stamp program, one of the pet issues of the consumer movement. Organized labor's quid pro quo was farm bloc endorsement of a higher minimum wage, food stamps for strikers, and a special minimum wage for sugar field workers (Paalberg, 1980: 29-30).

In 1977 the defusing of what had been intense political conflict over farm and food policies was a result of the political acumen of the groups involved. Larger commodity producers in particular, who had come to understand that the farm bloc had become isolated and could no longer write its own ticket, had adapted by reaching out for urban-based allies. Labor and consumer groups were receptive to these overtures by 1977, since the government had developed mechanisms for ensuring that exports would not unduly affect domestic food prices or supplies. With the priority on domestic food concerns clearly established, subsidiary policy differences were downplayed so that the groups could pursue other goals.

The perhaps inevitable resolution of the food crisis in favor of domestic interests left the foreign policy makers (and thus indirectly the LDCs) trying to salvage at least some influence in the decision-making process. The food crisis was a double-barrelled blow to the foreign policy community. First, it pulled the rug out for under one of the fundamental assumptions of the development paradigm: the continued availability of surplus commodities as a bridge until LDCs achieved food self-sufficiency. Second, official (USDA, 1980: 9) discovered that food had become a more important foreign policy resource than they had realized, both in offsetting growing U.S. balance of payments deficit and in substituting for dollars in foreign aid programs. All of these issues engendered controversy as the search for a new American foreign agricultural policy proceeded.

Among the first indicators of conflict was Agriculture Secretary Butz's omission of funding for PL 480 in his FY 1975 budget. Destler (1980: 68) explains that when asked about this by OMB, Butz responded "If Henry needs it, then let the money come out of his budget!" The struggle between the State and Agriculture Departments over the survival of PL 480 involved policy priorities (domestic versus foreign), bureaucratic power (State versus Agriculture), and personal influence (Kissinger versus Butz). State Department interests eventually prevailed in the struggle, but not without first having to appease both humanitarians and Congressional liberals on the question of criteria for the allocation of foreign aid.

Essentially the State Department had a foreign policy resource problem. Foreign aid programs traditionally had lacked a strong domestic constituency, and with inflation fueled by deficit spending, these programs become vulnerable to the wielders of the budget-cutting axe. Although the dollar amount of bilateral aid had increased slowly throughout the 1960s and 1970s, inflation and devaluation had relentlessly eroded its buying power. Congress was also reluctant to give the Johnson and Nixon administrations a blank check for the conduct of the Vietnam War, which led to budget and resource juggling at State as a means of continuing the war. Among other things, this involved a reallocation of PL 480 aid such that "by 1973, Title I food aid (low interest credit sales) for the hungriest nations had virtually disappeared; the U.S. gave priority instead of political purposes. In fiscal year 1973, South Vietnam and South Korea alone received almost half of all Title I food aid; in 1974 South Vietnam and Cambodia alone received 69 percent (Destler, 1980: 66). Thus, precisely at the same time that food scarcities were developing globally and food aid was most crucial, the U.S. had decided to funnel the bulk of what was available to its Indochina misadventure.

That proved to be an error of judgement, one which moved Congress to enact the 1975 amendment to PL 480 (called PL 94-161, the International Development and Food Assistance Act of 1975) and restrictively delineated eligibility requirements for Title I sales. Henceforth, "the law required that at least 75 percent of Title I sales go to countries with an annual per capita gross national product of $300 or less, and affected through their own production or through commercial purchase from abroad." (USDA, 1977: 4). In addition, reflecting liberal concern for what Senator Mark Hartfield termed a policy in which "preserving puppet regimes is more important than ... preserving the lives of millions of people [Congress also disallowed Title I agreements with countries that engage] in a consistent pattern of human rights violations" (USDA, 1977: 78). In other words the State Department had succeeded in saving PL

480, but because of its (and particularly Henry Kissinger's) role in Vietnam War, it paid the price of having its freedom to allocate PL 480 resources severely circumscribed by Congress.

After 1975, both international and domestic conditions had stabilized enough to allow a more measured, rational approach to foreign agricultural policy. Without the pressures of day-to-day crisis management, policy-makers were able to take stock of the new international economic environment and it implications for American global interests. With the priority on ensuring an adequate, low price domestic food supply clearly established, the other pieces of the foreign agricultural policy matrix could be constructed. Bureaucratic and interest group conflicts continued to exist, but they became more muted once the uncertainties of the crisis period had been resolved.

Important changes had occurred in the international sphere which affected the possibilities for U.S. action. Global recession and runaway inflation were destabilizing established trade patterns and severely straining the anti-Soviet alliance system among the industrialized Western nations. Periodic economic summitry had succeeded in papering over the cracks in the alliance without removing the underlying sources of conflict. Issues of contention included access to oil supplies, policy toward the Soviets, coordination of monetary policies, levels of defense spending, deployment of nuclear weapons, trade and tariff policies, and levels of aid-giving.

In addition the LDCs had become more united and vocally anti-American as a result of world economic crisis whose effects on them had been devestating. Their polite requests of the 1960s for more advantageous terms of trade became, in the late 1970s, demands for the creation of a New International Economic Order (NIEO) in which they would have power commensurate with their populations and resources. Perhaps emboldened by America's humiliation in Indochina, they increasingly used international forums such as the United Nations and regional organizations to press their demands. Ans lacking other types of leverage, the LDCs attempted to turn weakness into strength by threatening to default on their massive debt obligations unless the Western nations (and the international lending institutions which they controlled) agreed to restructure the debts and revise the terms of aid-giving.

The primary American response to the new situation was endorsement of the basic human needs approach to development in the LDCs, outlined in the 1974 World Bank study (Chenery,

1974). This report was a remarkably frank acknowledgement of the failures of the first two and a half decades of development theories and policies. In particular, it established that overall economic growth rates in the LDCs were being fettered by the existence of large numbers of poverty-stricken people. Those living in absolute poverty, often outside the monetarized economy, were unable to contribute to measurable economic activity. They were consumers of scarce resources through social service programs, but the resources expended in such programs were inadequate to enable the poor to translate their consumption into a widening, self-sustaining spiral of income improvement. Like American welfare recipients, the absolutely poor in the LDCs were locked into a situation from which there was no escape: they were a drain on the economy, because the investment required to make them economically productive was not forthcoming. In a following report the World Bank advised the following:

> "Some redirection of the investment of the economy toward the poverty groups can modify this process substantially over one or two decades. If it is provided in an appropriate mix of education, public facilities, access to credit, land reform, and so forth, investment in the poor can produce benefits in the form of higher productivity and wages in the organized sectors as well as greater output and income for the self-employed poor. In the shortrun, there may be a reduction in the growth of other groups through this redirection of investment toward the poor, although it is by no means necessary. In the longer run, however, it can be argued that the transformation of poverty groups into more productive members of society is likely to raise the income of all" (IBRD, 1979: 13).

For humanitarian and ultimately political reasons, the economic disparities between rich and poor in the LDCs were no longer deemed acceptable outcomes of the development process. Political scientists had accumulated data demonstrating a strong correlation between equitable income distribution and that most crucial American foreign policy objective, political stability in the LDCs (Mellor, 1966). Hence the elimination of absolute poverty was also posted as an alterative or deterrent to future chaos in the international system.

The linkages between the basic human needs strategy and U.S. foreign agricultural policy were clearly established by the World Bank. It is explained (World Bank, 1975) that "Of the population in the developing countries considered to be either absolute or relative poverty, more than 80 percent are estimated to live in rural areas. Agriculture is the principal occupation of four-fifths of the rural poor." With poverty so highly concentrated in the agricultural sector, the most obvious approach to eradicating poverty was to focus resources on programs of rural development which would be clearly designed to increase production and raise productivity. However, given the scope of the problem, its solution was beyond the resources of a single donor country. As envisioned by the World Bank, the attempt to upgrade rural living standards and increase agricultural productivity in the LDCs required a massive infusion of external aid as well as careful coordination of bilateral and multilateral programs.

In response, the U.S. shifted its emphasis in aid-giving from bilateralism to multilateralism. Multilateralism was attractive, albeit for different reasons, to both the foreign policy makers and the LDCs. For the foreign policy community there were three perceived advantages. First, increased multilateralism involved no real loss of control over aid-giving in (in terms of goals, eligibility criteria and allocation decisions). Voting rights in the multilateral institutions such as the World Bank and the IMF were weighted to favor the large contributors, and the U.S. was the largest contributor (Destler, 1980: 47). Second, it provided the policy makers a means to circumvent Congressional restrictions on aid-giving, especially those which disqualified recipients with records of human rights abuses. Congressional efforts to close this loophole faltered when the implications of any country's ability to dictate policy to multilateral institutions were understood. Finally, only a collective approach could generate sufficient resources to begin solving the LDCs massive agrarian problems. Neither private investors or individual donor nations were willing to undertake large, risky, or experimental programs with long payback times. However, multilaterals such as the International Development Association (IDA), a World Bank affiliate created to provide virtually interest-free loans to the very poorest LDCs, could fill this need. By 1978 IDA had provided nearly $5 billion for agricultural and rural development including project in Burma for seed development, in Nepal for improving irrigation systems, and in Sri Lanka for upgrading the tea industry (World Bank, 1978: 72-81, 176-177).

As for aid recipients, multilateralism had the attraction of providing assistance from a relatively source. LDCs objected to bilateral aid programs, in particular those sponsored by the U.S. because they frequently had strings and/or implicit political obligations attached. The strings could take the form of the carrot, as with PL 480 aid to Egypt in return for its agreement to make a separate peace with Israel, or the stick as with the threat with withhold aid to India if it persisted in its development of nuclear weapons. And the political manipulation of foreign aid by its donors meant that assistance was often not available to those who needed it most. USAID's FY 1980 budget request is illustrative. AID requested $1,801,879,000 in total bilateral development assistance and $1,995,100,000 for security supporting assistance (SSA), defined as "economic assistance to countries based on considerations of special, political or security needs and U.S. interests" (AID, 1980: 22, 206). All but $148.1 million or 86 percent, of the nearly $2 billion budgeted for SSA went to the Middle East, with the lion's share going to Israel ($785 million) and Egypt ($750 million) "as a part of continuing efforts to achieve a stable peace" (AID, 1980: 30). In short, the total amount requested for SSA (political purposes) exceeded that requested for development purposes three years after Congress had specified that the bulk of U.S. foreign assistance must go the poorest, neediest countries. Of the $1.8 billion bilateral aid nutrition programs for the "seven hundred million people---almost one-fifth of humanity---who were seriously malnourished" (AID, 1980: 34-35). Thus, despite the apparent American commitment to the basic human needs strategy, policy makers could not resist the temptation of using bilateral aid for immediate political ends.

Achieving the objectives of eliminating absolute poverty globally by the year 2000 required a fairly complete reorientation of the U.S. foreign agricultural policy-making process. American domestic agricultural production had to be planned with an eye to its international implications. Given the competition which had arisen between commercial and concessional transactions during the period of scarcity, such planning had to ensure the continued availability of adequate supplies of food aid until LDC agriculture became more self-sufficient. In addition, U.S. participation in the proposed international grain reserve program was vital to insulate the nutritional goals of the basic human needs strategy from short-term fluctuations in the world food supply (U.S. House, 1984). American bilateral assistance programs---both PL 480 and functional development aid---had to be reformulated, adequately funded, and integrated with multilateral assistance efforts. Finally, the U.S. emphasis on multilateralism required a sustained, large-scale commitment of resources free of political and ideological interference.

THE REAGAN FACTOR

Despite the above affects of reorienting the U.S. foreign agricultural policies very few of these changes had occurred by the mid-1980s, largely as a result of the Reagan victory in the 1980 Presidential election. Although the Reagan administration, of necessity, has made international economic policy a higher priority than its predecessors, its anti-interventionist, laissez faire economic philosophy has precluded the initiatives needed to create a coherent, reciprocally beneficial foreign agricultural policy. Rather than reaffirming the U.S. commitment to the basic human needs strategy, Reagan has politicized all aspects of aid-giving and has adopted the same victim-blaming attitude toward the poverty in the LDCs that he has enunciated toward the American poor. In essence the Reagan administration has repudiated the lessons of the previous fifteen years of both economic development and foreign agricultural policy experience, leaving the policy process in disarray and the LDCs to fend for themselves.

The Reagan approach to development in the LDCs has been called "private sector voodoo" (<u>Economist</u>, 31 October 1981: 34) by its opponents and encouragement of private sector initiative by its supporters (Madison, 1982). Its proponents insist that the fundamental propose of development programs---relieving absolute poverty---has not changed; instead there has simply been a shift of emphasis away for direct aid to the poor and toward working "with host governments to modify their economic policies in order to increase the climate for private-sector investment" (Madison, 1982: 963). They argue that they key to development is increasing productivity, and that this can be achieved most efficiently through free markets and private sector initiatives.

This pro-capitalist politicization of foreign aid has meant, among other things, reduced support for integrated rural development projects in LDCs where "appropriate" policy changes have not occurred. Other ideologically inspired restrictions, as well, have characterized the foreign aid segment of the U.S. foreign agricultural policy during the Reagan administration. Nations such as Zimbabwe, which publicly criticize America or American policies, are systematically declared ineligible for bilateral assistance regardless of need. And the revival of Cold War anti-communist rhetoric accompanied by aggressively confrontational American behavior has meant that any LDC that does not respond positively to U.S. demands, or that has been tagged with the "Marxist" label, or that is of no strategic importance to the United States is unlikely to receive direct aid, again regardless of need.

At least until 1982, Congress resisted AID's attempt to reshape bilateral aid programs. During its lame duck sitting in December, 1982, it passed amendments to the 1961 Foreign Assistance Act that required a minimum of 40 percent of FY 1983 foreign aid, and 50 percent of FY 1984 foreign aid to be used to "expeditiously and directly benefit those living in absolute poverty..." (U.S. House, 1982). The amendments specifically mentioned rural development projects---irrigation, extension services, roads, agricultural credit, etc., as being appropriate uses for foreign assistance funds. The more conservative 98th and 99th Congresses, however, have been more amenable to the Reagan administration's approach to foreign aid and have moved toward a position of making friendly nation status (defined ideologically and strategically) more important than need in the allocation of bilateral assistance.

Both Reagan and the Congress have undermined the principles of multilateralism since 1980 by threatening withdrawal from multilateral organizations hostile to American policies and refusing to continue funding to programs, even parts of which fail to pass ideological muster. Probably the most prominent instance of the latter was American refusal to contribute to multilateral health programs when even a small part of the services they provided included abortions. There has also been less willingness on the part of the United States to contribute directly to multilateral development programs; instead as Feder (1986: 18) points out American policy makers have proposed substituting commercial financing for large segments of such projects, even though economists doubt that private financial institutions would be forthcoming with the large amounts of capital required.

Domestically, the re-emergence of commodity gluts in the 1980s has rekindled farmer concern about both farm and foreign agricultural policies. Reagan's proposed elimination of farm programs and a shift to reliance solely on market forces were strongly resisted by the farm bloc and its Congressional supporters. This attempt to wean agriculture from government intervention, combined with the tightening of rural credit and a rising rate of farm foreclosures in the Midwest, may seriously erode support the Republican Party in agricultural states.

Farming interests have also been concerned about the international strength of the dollar, which has made commodity exports expensive, has reduced the volume of such exports, and has contributed to the record-breaking trade deficits of the past several years. Revaluation of the dollar has failed to stimulate commercial commodity transactions in last part because the wealthier LDCs like Brazil have become net commodity exporters themselves ruing this period. Don Paarlberg (1980: 251) was prescient in 1980 when he predicted the U.S. "will be looking for places to put food rather than withholding it as a diplomatic lever. The importing nations may consider that they do us a favor by accepting it."

The lack of commercial demand for American commodities has not been accompanied by lack of concessional demands or need for disaster relief, as evidenced in particular by the sub-Saharan famines. The Reagan administration's combination of free market and Cold War ideologies were quite evident in its response to the Ethiopian famine. From this perspective the famine was a result not of uncontrolled population growth, drought, and progressive environmental degradation, but rather a consequence of Marxist agricultural and economic policies. Since Ethiopia was an unfriendly country with a non-capitalist economic system, American disaster relief was slow to arrive and reluctantly given despite more-than-adequate surplus stocks. The massive private fundraising and relief efforts through Food Aid and other channels have proved more effective than American government contributions toward meeting the Sahel's long-term food needs. Unfortunately such volunteerism is a response to a specific crisis and cannot be sustained, and quantitatively it is no substitute for government-to-government and multilateral development assistance.

Therefore, in 1987 the United States still lacks a coherent foreign agricultural policy. The possibilities for constructing such a policy were present at the beginning of the decade with the apparent U.S. commitment to the basic human needs strategy. But the Reagan administration has rearranged American priorities and their ideological underpinnings both domestically and internationally. The absence of consensus on domestic farm policy and administration antipathy to government intervention in the economy make it difficult to foresee the implementation of production planning or coordination with multilateral bodies. Commodity production will continue to fluctuate without concern for the state of global grain reserves or the disincentive effects of surplus disposal on the LDCs[3]. In foreign policy terms, the politicization of aid programs and the attempt to substitute private investment for multilateral assistance is unlikely to solve the food and agricultural problems of the LDCs. Rather it can potentially backfire on the U.S. by exacerbating inequalities, preventing necessary structural change, and generating the political instability that policy-makers have tried for decades to prevent.

CONCLUSION

The study of U.S. foreign agricultural policy making is the investigation of a messy, uncoordinated, conflictive process. Its content is determined by the interplay of domestic and international, political and economic, practical and ideological factors which are constantly being reordered as conditions change. Furthermore, foreign agricultural policy since World War II has not been an autonomous policy area. Its linkages to broader foreign policy objectives and to domestic economic policies have insured that its content is always the result of compromise.

As a subset of foreign policy in general and foreign economic policy in particular, it must conform to the parameters of these areas which provide the structure for its formulations and implementation. Foreign economic policy itself has tended to be neglected and uncoordinated in the postwar period except during period of crisis. Most of the attention it has received has focused on the maintenance of international economic institutions rather than actual substance of policy or the policy-making process. For foreign agricultural policy the result has been an amorphorous, shifting set of guidelines within which to operate, accompanied by episodic Executive or Legislative branch intervention.

Domestic political considerations have further complicated the situation. The farm bloc's vested interest in tariff barriers, in export and production subsidies, and in maximizing commercial exports have received domestic political support even though they often produce a foreign agricultural policy in conflict with prevailing foreign policy goals such as freer trade or greater food self-sufficiency abroad. Bureaucratic infighting, the vagaries of the economy, and the discontinuities inherent in the American electoral process also contribute little to coherence and rationality in the making of foreign agricultural policy.

However, with all these contradictory influences pulling in opposite directions, U.S. foreign agricultural policy has served to extend and preserve American global interests. It has done this by contributing to the creation of an international economic interdependence so effective that leaders like Chile's Allende, who choose to opt out, faced severe food shortages and political unrest when food aid was discontinued. It has done so when food has served as an enticement for other nations to modify their international behavior in ways beneficial to the U.S., as with the Egyptian-Israeli peace accords. And it has provided the United States an entree into the LDCs, either directly through bilateral aid programs, or indirectly through its multinational corporations, control of multilateral institutions, and participation in multilateral assistance efforts aimed at economic and agricultural development.

NOTES

1. The total amount of aid to agriculture in 1971 was approximately $263 million; PL 480 long-term dollar sales were $539 million. The $300 million addition to the debt burden (subtracting aid from sales) is a minimum figure since at least a part of the aid to agriculture was in the form of loans repayable in dollars, not grants.

2. Don Paarlberg, Farm and Food Policy: Issue of the 1980s,(Lincoln University of Nebraska Press: 1980, Chapter 7). Consumers demanded the end of tobacco subsidies, the banning of hazardous pesticides (DDT) and substances (DES), nutritional content labeling and unit pricing.

3. For detailed discussion of this matter see: Disincentives to Agricultural Production in Developing Countries, Report to Congress of the Comptroller General of the United States, ID 76-2, November 26, 1975; Abdullah A. Saleh, Disincentives to Agricultural Production in Developing Countries", Foreign Agriculture, FAS, USDA, Supplement March, 1975; Frank D. Barlow and Susan A. Libbin, Food Aid and Agricultural Development, Economic Research

Service, USDA, Foreign Agricultural Economic Report, No. 51, 1969, Paul J. Iseman and H.W. Singer, "Food Aid: Disincentive Effects and their Policy Implications", Economic Development and Cultural Change, July 1976.

REFERENCES

Agency for International Development (1980). Congressional Presentation: Fiscal Year. Washington, D.C.

Allen, Rau (1957). Agricultural Policy and Trade Liberalization in the United States, 1934-1956: A Study of Conflicting Policies. Geneva: E. Droz.

Barlow, Frank D. and Susan A. Libben (1969). Food Aid and Agricultural Development. Washington: USDA.

Chenery, Hollis, et al. (1974). Redistribution With Growth: Policies to Improve Income Distribution in the Context of Economic Growth. Sussex: Oxford University Press.

Cockfort, James D., A.G. Frank, and Dale L. Johnson. eds. (1972). Dependence and Underdevelopment: Latin America's Political Economy. Garden City: Doubleday & Co.

Destler, I.M. (1980). Making Foreign Economic Policy. Washington: Brookings Institution.

The Economist. Vol. 281, no 7209 (31 October 1981).

Feder, Barnaby J. "New Study Offers Plan to Rescue Latin Economies" New York Times (28 September 1986).

Hickman, C. Addison (1949). Our Form Program and Foreign Trade: A Conflict of National Policies. New York: Council on Foreign Relations.

Isenman, Paul J. and H.W. Singer (1979). "Food Aid: Disincentive Effects and Their Policy Implications." Economic Development and Cultural Change (July).

Madison, Christopher. "Exporting Reagonomics-The President Wants to Do Things Differently at AID" National Journal (29 May 1982).

Mellor, John W. (1966). The Economics of Agricultural Development. Ithaca: Cornell University Press.

Paarlberg, Don (1980). Farm and Food Policy: Issues of the 1980s. Lincoln: University of Nebraska Press.

Peterson, Trudy Huskamp (1979). Agricultural Exports, Farm Income, and the Eisenhower Administrations. Lincoln: University of Nebraska Press.

Rostow, W.W. (1960). The Stages of Economic Growth: A Non-Communist Manifesto. Cambridge: Cambridge University Press.

Saleh, Abdullah (1975). "Disincentives to Agricultural Production in Developing Countries". Foreign Agriculture. Washington: USDA.

Saylor, Thomas Reese (1977). "A New Legislative Mandate for American Food Aid." In G. Brown and Henry Shue. eds. Food Policy: The Responsibility of the United States in the Life and Death Choices. New York: Free Press.

U.S. President's Office (1966). FY 1966 President's Report on the Foreign Assistance Program. Washington: Government's Printing Office.

VanGigch, Francis (1972). "Historical and Economic Summary of U.S. Agriculture." In Glenn L. Johnson and C. Leroy Quance. eds. The Overproduction Trap in U.S. Agriculture A Study of Resource Allocations from World War I to the Late 1960s. Baltimore: The John's Hopkins University Press.

World Bank (1975). The Assault on World Poverty: Problems of Rural Development, Education and Health. Baltimore: The Johns Hopkins University Press.

_____ (1978). World Bank Annual Report, 1978. Washington: The World Bank.

_____ (1979). World Development Report, 1979. New York: Oxford University Press.

Yost, Israel (1971). "The Food for Peace Arsenal". <u>NACLA Newsletter</u>, V, no. 3 (May-June).

Contributors

VALERIE J. ASSETTO is Assistant Professor of Political Science at Colorado State University in Fort Collins. She is the author of The Soviet Bloc in the IMF and IBRD (Westview Press, 1987) and her current research interests are international assistance, politics of agriculture, international organizations, and Eastern European politics. She is a recipient of the IREX Fellowship in 1979. She received her Ph.D. in 1984 from Rice University.

CHARLES D. BROCKETT is Associate Professor of Political Science at the University of the South in Sewanee, Tennessee. He is the author of Land, Power and Politics: Agrarian Transformation and Political Conflict in Central America (forthcoming). He has published numerous articles on Guatemala, Honduras, and international distributive justice. He received his Ph.D. in 1974 from the University of North Carolina at Chapel Hill.

GRETCHEN CASPER is Assistant Professor of Political Science at Texas A & M University. She is currently studying patterns of regime instability under authoritarianism in Philippines, particularly during the Marcos era. She received her Ph.D. in 1986 from the University of Michigan at Ann Arbor.

RONALD COX is a Ph.D. student of Political Science at the University of Wisconsin in Madison. His research interests include radical political economy, Central America, and North-South relations.

JEAN DOYLE is Professor of Political Science and Assistant to the Dean of Arts and Sciences at the Southeastern

Massachusetts University. Her research interests and publications have been in the areas of international relations/world politics and comparative politics of communist systems. She received her Ph.D. in 1973 from Boston University.

BRUCE R. DRURY is Professor of Political Science at Lamar University in Texas. His major research interests and publications have been in the area of civil-military relations and Brazilian public policy. His present research interest is agricultural policy in developing countries. During the 1986-1987 academic year he was on a research leave from Lamar University and was associated with the MARA Technical Institute of Malaysia. He received his Ph.D. in 1973 from the University of Florida.

MAHIR FISUNOĞLU is Associate Professor of Economics at Cukurova University in Adana, Turkey. His major research interests and publications have been in the areas of general equilibrium dynamics, macroeconomic modelling for developing countries, and economic development in Turkey. He received his Ph.D. in 1983 from the University of Michigan at Ann Arbor.

GARY HAWES is Assistant Professor of Political Science at the University of Michigan. He is the author of The Philippine State and the Marcos Regime: The Politics of Export (Cornell University Press, 1987) and is currently researching the impact of the Marcos regime on the Philippine political system. He received his Ph.D. in 1984 from the University of Hawaii.

CYNTHIA McCLINTOCK is Associate Professor of Political Science at George Washington University. She is the author of Peasant Cooperatives and Political Change in Peru (Princeton University Press, 1981) and the co-editor of The Peruvian Experiment Reconsidered (Princeton University Press, 1983). During the 1980s, she has returned frequently to the same agricultural cooperatives and villages in Peru that she first studied in the early 1970s. She carried out research on Ecuadorian agricultural policies in the summers of 1985 and 1986. She received her Ph.D. in 1976 from the Massachusetts Institute of Technology.

WAYNE MOYER is Rosenfield Professor, Professor of Political Science, and Director of the Rosenfield Program in Public Affairs, International Relations and Human Rights at

Grinnell College in Iowa. His major research and publications have been in the area of international organizations and law, world politics, and food policy. He received his Ph.D. in 1976 from Yale University.

ADALBERTO J. PINELO is Professor and Chairperson of Political Science at Northern Kentucky University. He specializes in Latin American Politics. He is a consultant in Political Risk Analysis and is the author of The Multinational Corporation as a Force in Latin American Politics: A Case Study of the International Petroleum Company in Peru (Praeger, 1973) and World Political Risk Forecast: Argentina (Frost and Sullivan, 1981). He received his Ph.D. in 1972 from the University of Massachusetts at Amherst.

ROSS B. TALBOT is Professor of Political Science at Iowa State University. His teaching in part, and his research, almost exclusively, is concerned with world food and development assistance politics and policies. He is the author of numerous works on these subjects. He received his Ph.D. in 1953 from the University of Chicago.

TAYE WOLDESMIATE is a Ph.D. candidate in Political Science at the University of Missouri-Columbia. His research interests include African politics, agriculture in the Third World, and radical political economy.

BIROL A. YEŞILADA is Assistant Professor of Political Science at the University of Missouri-Columbia. His major research interests and publications have been in the areas of North-South relations, international political economy, and Turkish politics. He received the Fulbright-Hays Research Fellowship in 1982 and a grant from the Joint Committee on Near and Middle East of the American Council of Learned Societies and the Social Science Research Council in 1987. He is currently on a research leave from the University of Missouri-Columbia and is associated with the Middle East Technical University in Ankara, Turkey. He received his Ph.D. in 1984 from the University of Michigan at Ann Arbor.

DATE DUE

APR 15 1992			
APR -6 1994			
	APR 7 1994		
JUN 19 1994			
MAY 26 1994			
	261-2500		Printed in USA